Smart Water Management

This book contributes to the debate about the suitability and challenges of the Smart Water Management (SWM) approach. Smart Water Management has increasingly been promoted to manage water and wastewater more efficiently and cost effectively by industries and utilities in urban contexts at regional or city scales, while reducing overall consumption. It is based on the use of Information and Communication Technology (ICT) to provide real-time, automated data to resolve water challenges. Many of these technologies are complex and costly, however, and the approach tends to overlook cheaper and less high-tech (softer) approaches to address the same problems. Yet there may be opportunities for using them even in resource short rural communities in developing countries.

The book includes examples of SWM systems in practice in diverse locations from Korea, Mexico, Paris, the Canary Islands and southern Africa, aimed at addressing a diverse set of problems, including monitoring water supply to refugees. Critical voices highlight the need for smart institutions to accompany smart technologies, the absurdity of applying SWM to dysfunctional legacy infrastructure systems whether its adoption raises moral hazards or SWM is the latest example of hegemonic masculinity in water management.

The chapters in this book were originally published in *Water International.*

Henning Bjornlund is a Fellow and Vice President at IWRA. He is also a Research Professor in Water Management and Policy at the University of South Australia. He has been researching water management and policy issues in Australia since 1993, in Canada since 2005 and in southern Africa since 2013 where he works closely with partners in Mozambique, Tanzania, Zimbabwe, South Africa and Ethiopia.

Stephanie Kuisma, a former Project Consultant for IWRA, is a climate and sustainability specialist in Melbourne, Victoria, Australia.

James E. Nickum, Fellow, Global Reach Awardee and former Vice-President of the International Water Resources Association (IWRA) is the Editor in Chief of *Water International*, non-resident Professorial Research Associate at the School of Oriental and African Studies (SOAS), University of London and non-resident Adjunct Senior Research Fellow, East-West Center, Honolulu.

Raya Marina Stephan is Fellow and Former Director of IWRA. She is an expert in water law, and an international consultant in water related projects with international organizations. She is the Deputy Editor in Chief of *Water International.*

Routledge Special Issues on Water Policy and Governance

Edited by: Cecilia Tortajada *(IJWRD) – Institute of Water Policy, Lee Kuan Yew School of Public Policy, NUS, Singapore,* and **James E. Nickum** *(WI) – International Water Resources Association, France*

Most of the world's water problems, and their solutions, are directly related to policies and governance, both specific to water and in general. Two of the world's leading journals in this area, the *International Journal of Water Resources Development* and *Water International* (the official journal of the International Water Resources Association), contribute to this special issues series, aimed at disseminating new knowledge on the policy and governance of water resources to a very broad and diverse readership all over the world. The series should be of direct interest to all policy makers, professionals and lay readers concerned with obtaining the latest perspectives on addressing the world's many water issues.

Global Water Resources
Festschrift in Honour of Asit K. Biswas
Edited by Cecilia Tortajada and Eduardo Araral

Source-to-Sea Management
Edited by Josh Weinberg, Qinhua Fang, Sarantuyaa Zandaryaa, Greg Leslie and James E. Nickum

Strengthening Cooperation over Transboundary Groundwater Resources
Edited by Gabriel Eckstein and Alice Aureli

Wicked Problems of Water Quality Governance
Edited by James E. Nickum, Raya Marina Stephan and Henning Bjornlund

Groundwater
Recent Advances in Interdisciplinary Knowledge
Edited by Raya Marina Stephan, James E. Nickum and Philippus Wester

Public Banks, Public Water
Exploring the Links in Europe
Edited by Thomas Marois and David A McDonald

Smart Water Management
Truly Intelligent or Just Another Pretty Name?
Edited by Henning Bjornlund, Stephanie Kuisma, James E. Nickum and Raya Marina Stephan

For more information about this series, please visit:
www.routledge.com/Routledge-Special-Issues-on-Water-Policy-and-Governance/book-series/WATER

Smart Water Management

Truly Intelligent or Just Another Pretty Name?

Edited by

Henning Bjornlund, Stephanie Kuisma, James E. Nickum and Raya Marina Stephan

Routledge
Taylor & Francis Group
LONDON AND NEW YORK

IWRA

First published 2023
by Routledge
4 Park Square, Milton Park, Abingdon, Oxon OX14 4RN

and by Routledge
605 Third Avenue, New York, NY 10158

Routledge is an imprint of the Taylor & Francis Group, an informa business

British Library Cataloguing in Publication Data
A catalogue record for this book is available from the British Library

ISBN13: 978-1-032-34573-4 (hbk)
ISBN13: 978-1-032-34574-1 (pbk)
ISBN13: 978-1-003-32286-3 (ebk)

DOI: 10.4324/9781003322863

Typeset in Minion Pro
by Newgen Publishing UK

Publisher's Note
The publisher accepts responsibility for any inconsistencies that may have arisen during the conversion of this book from journal articles to book chapters, namely the inclusion of journal terminology.

Contents

PART III
Innovative Uses and Critical Perspectives

Citation Information

The chapters in this book were originally published in various volumes and issues of the journal *Water International*. When citing this material, please use the original page numbering for each article, as follows:

Chapter 1

Smart Water Management: the way to (artificially) intelligent water management, or just another pretty name?
James E. Nickum, Stephanie Kuisma, Henning Bjornlund and Raya Marina Stephan
Water International, volume 45, issue 6 (2020), pp. 515–519

Chapter 2

The report that sparked this special issue
Stephanie Kuisma, Callum Clench, Monica Garcia Quesada, James E. Nickum and Henning Bjornlund
Water International, volume 45, issue 6 (2020), pp. 520–525

Chapter 3

SWM technology for efficient water management in universities: the case of PUMAGUA, UNAM, Mexico City
Fernando González Villarreal, Cecilia Lartigue, Josué Hidalgo, Berenice Hernández and Stephanie Espinosa
Water International, volume 45, issue 6 (2020), pp. 526–551

Chapter 4

K-water's Integrated Water Resources Management system (K-HIT, K-water Hydro Intelligent Toolkit)
Sukuk Yi, Munhyun Ryu, Jinsuhk Suh, Shangmoon Kim, Seokkyu Seo, Seonghan Kim and Sungphil Jang
Water International, volume 45, issue 6 (2020), pp. 552–573

Chapter 5

Integrated Smart Water Management of the sanitation system of the Greater Paris region
Jean-Pierre Tabuchi, Béatrice Blanchet and Vincent Rocher
Water International, volume 45, issue 6 (2020), pp. 574–603

Chapter 6

Is Smart Water Management really smart? What experts tell us
James E. Nickum, Henning Bjornlund, Raya Marina Stephan and Stephanie Kuisma
Water International, volume 45, issue 6 (2020), pp. 604–607

Chapter 7

Smart water management: can it improve accessibility and affordability of water for everyone?
Neil S. Grigg
Water International, volume 45, issue 6 (2020), pp. 608–620

Chapter 8

Institutional innovation and smart water management technologies in small-scale irrigation schemes in southern Africa
H. Bjornlund, A. van Rooyen, J. Pittock, K. Parry, M. Moyo, M. Mdemu and W. de Sousa
Water International, volume 45, issue 6 (2020), pp. 621–650

Chapter 9

Using innovative smart water management technologies to monitor water provision to refugees
Ryan W. Schweitzer, Ben Harvey and Murray Burt
Water International, volume 45, issue 6 (2020), pp. 651–659

Chapter 10

A GIS-based solution for urban water management
Pablo Fernández Moniz, Jaisiel Santana Almeida, Agustín Trujillo Pino and José Pablo Suárez Rivero
Water International, volume 45, issue 6 (2020), pp. 660–677

Chapter 11

SWM and urban water: Smart management for an absurd system?
M. P. Trudeau
Water International, volume 45, issue 6 (2020), pp. 678–692

Chapter 12

The moral hazards of smart water management
Kris Hartley and Glen Kuecker
Water International, volume 45, issue 6 (2020), pp. 693–701

Chapter 13

Masculinity and smart water management: why we need a critical perspective
Anna Kosovac
Water International, volume 46, issue 3 (2021), pp. 342–344

Chapter 14

Deconstructing masculinity in water governance
Kris Hartley and Glen Kuecker
Water International, volume 46, issue 5 (2021), pp. 671–676

Conclusion

Before you go: the editors' checklist of what we now know about Smart Water Management
Stephanie Kuisma, James E. Nickum, Henning Bjornlund and Raya Marina Stephan
Water International, volume 45, issue 6 (2020), pp. 702–703

Notes on Contributors

Henning Bjornlund, UniSA Business School, University of South Australia, Adelaide.

Béatrice Blanchet, SIAAP (Syndicat interdépartemental pour l'assainissement de l'agglomération parisienne), Paris, France.

Murray Burt, Auckland Transport, Auckland, New Zealand.

Callum Clench, IWRA, Paris, France.

Stephanie Espinosa, Programme for the Management, Use and Reuse of Water (PUMAGUA), National Autonomous University of Mexico (UNAM), Mexico City, Mexico.

Pablo Fernández Moniz, Information and Communications System, Division of Mathematics, Graphics and Computation, Universidad de Las Palmas de Gran Canaria, Spain.

Fernando González Villarreal, Programme for the Management, Use and Reuse of Water (PUMAGUA), National Autonomous University of Mexico (UNAM), Mexico City, Mexico.

Neil S. Grigg, Civil and Environmental Engineering, Colorado State University, Fort Collins, USA.

Kris Hartley, Department of Asian and Policy Studies, Education University of Hong Kong, Tai Po, China.

Ben Harvey, Division of Resilience and Solutions, United Nations High Commissioner for Refugees, Geneva, Switzerland.

Berenice Hernández, Programme for the Management, Use and Reuse of Water (PUMAGUA), National Autonomous University of Mexico (UNAM), Mexico City, Mexico.

Josué Hidalgo, Programme for the Management, Use and Reuse of Water (PUMAGUA), National Autonomous University of Mexico (UNAM), Mexico City, Mexico.

Sungphil Jang, K-water (Korea Water Resources Corporation), Daejeon, Korea.

Seonghan Kim, K-water (Korea Water Resources Corporation), Daejeon, Korea.

Shangmoon Kim, K-water (Korea Water Resources Corporation), Daejeon, Korea.

Anna Kosovac, Connected Cities Lab, The University of Melbourne, Melbourne, VIC, Australia.

Glen Kuecker, Department of History, DePauw University, Greencastle, IN, USA.

Stephanie Kuisma, International Water Resource Association (IWRA, former consultant), Melbourne, Victoria, Australia.

Cecilia Lartigue, Programme for the Management, Use and Reuse of Water (PUMAGUA), National Autonomous University of Mexico (UNAM), Mexico City, Mexico.

M. Mdemu, Institute of Human Settlement Studies, Ardhi University, Dar es Salaam, Tanzania.

M. Moyo, Innovation Systems for the Drylands, International Crops Research Institute for the Semi-Arid Tropics, Bulawayo, Zimbabwe.

James E. Nickum, IWRA, Tokyo, Japan.

K. Parry, UniSA Business School, University of South Australia, Adelaide, Australia.

J. Pittock, Fenner School of Environment & Society, The Australian National University, Acton ACT, Australia.

Monica Garcia Quesada, IWRA, Paris, France.

Vincent Rocher, SIAAP (Syndicat interdépartemental pour l'assainissement de l'agglomération parisienne), Paris, France

Munhyun Ryu, K-water (Korea Water Resources Corporation), Daejeon, Korea.

Jaisiel Santana Almeida, Information and Communications System, Division of Mathematics, Graphics and Computation, Universidad de Las Palmas de Gran Canaria, Spain.

Ryan W. Schweitzer, Division of Resilience and Solutions, United Nations High Commissioner for Refugees, Geneva, Switzerland.

Seokkyu Seo, K-water (Korea Water Resources Corporation), Daejeon, Korea.

W. de Sousa, National Irrigation Institute, Ministry of Agriculture and Food Security, Maputo, Mozambique.

Raya Marina Stephan, Water Law Expert, Consultant; Deputy Editor in Chief, Water International, IWRA, Paris, France.

José Pablo Suárez Rivero, Information and Communications System, Division of Mathematics, Graphics and Computation, Universidad de Las Palmas de Gran Canaria, Spain.

Jinsuhk Suh, K-water (Korea Water Resources Corporation), Daejeon, Korea.

Jean-Pierre Tabuchi, SIAAP (Syndicat interdépartemental pour l'assainissement de l'agglomération parisienne), Paris, France.

M. P. Trudeau, Envirings Inc., Ottawa, Ontario, Canada.

Agustín Trujillo Pino, Imaging Technology Center, Universidad de Las Palmas de Gran Canaria, Spain.

A. van Rooyen, Innovation Systems for the Drylands, International Crops Research Institute for the Semi-Arid Tropics, Bulawayo, Zimbabwe.

Sukuk Yi, K-water (Korea Water Resources Corporation), Daejeon, Korea.

Part I

Introducing Smart Water Management

Smart Water Management: the way to (artificially) intelligent water management, or just another pretty name?

James E. Nickum, Stephanie Kuisma, Henning Bjornlund
and Raya Marina Stephan

This is the second special issue prepared on behalf of the International Water Resources Association (IWRA) under the aegis of its Science, Technology and Publications Committee. The first, we hope you recall, appeared in April 2018 (vol. 41, no. 3): *The Wicked Problems of Water Governance*. These IWRA special issues are intended to complement existing IWRA initiatives.

As foreshadowed in the earlier special issue, the focus of this one is Smart Water Management (SWM). It evolved from a collaboration with K-water, the Korean water authority, that resulted in a report including a number of case studies of SWM from around the world. Following this introduction is a summary of the findings of that report, including a description of what SWM is (in a nutshell, the application of new technologies to water management), a general characterization of its progress to date, a summary of findings (benefits, factors for success and policy recommendations), and prospects for further development (Kuisma, Clench, et al., 2020).

We follow with a small sample of summaries of the cases in that collection, from Mexico (Villareal et al., 2020), Korea (Yi et al., 2020) and France (Tabuchi et al., 2020), covering a range of issues and domains, from water quality monitoring at a campus level, to a nationwide flood monitoring system, to the sanitation sector of a global city.

Villeareal et al. present a case of the development of SWM technology on a major university campus (UNAM) in Mexico. Here the project, focused on data monitors managed by a specific programme with the acronym PUMAGUA, has been aimed at improving water quality, reducing water consumption and encouraging behavioural change in students and the university community. This programme has reduced potable water consumption by 50% through improved practices and leak detection, while improving the quality of drinking water, treating wastewater to levels specified in Mexican regulations, and promoting participation of the entire university community in using water more efficiently. Yet the authors note that while SWM can build trust in water users, it requires a lot of time and effort to manage and maintain, especially in the early stages of its implementation. Given the speed and scope of technological change, they further argue that it is essential to have the financial resources and capacity to maintain and upgrade technology to ensure that is continues to yield benefits.

Yi et al. (2020) showcase the ability of SWM to significantly improve drought and flood management for an entire country (in this case, South Korea) through the use of an integrated SWM Hydro Intelligent Toolkit (K-HIT). By combining the use of real-time hydrological data acquisition, precipitation forecasting, flood analysis, reservoir water supply and hydropower generation, the authors demonstrate K-HIT's potential to minimize flood damage through the storage of water during the rainy season, which can then be used to prevent droughts during the dry season. To support successful large-scale implementation of SWM, Yi et al. advocate long-term investment, budgeting and support from the public sector. They further recommend proceeding step by step so as to adapt as necessary. Alongside water management benefits, the authors also highlight the increased capacity that SWM affords K-water to address water quality and reduce flood risk. SWM centred on K-HIT proved its worth by enabling K-water to effectively manage major floods in 2012, 2013 and 2015.

From Paris, Tabuchi et al. (2020) describe how a real-time control system (MAGES) was developed to better manage the Greater Paris sanitation system, with the search for optimal use of its treatment facilities, in particular for stormwater pollution caused by a combined sewer system. MAGES has been developed and implemented for over 20 years, the product of a strong collaborative process, with shared goals and strong commitment, the opportunity to develop advanced skills in hydraulics and urban hydrology, and to test ideas. The Greater Paris sewer system is decentralized; MAGES allows all to take into account the overall operation of the sanitation system, and to adapt the system to integrate many different systems, without merging all systems under a single administration. The system has greatly improved all the sanitation system management but will face significant challenges in the future, with probable reduction of the flow of the Seine due to climate change. Continuous improvements, enabled by technological advance, facilitating even smarter water management, are foreseeable and necessary.

After this global tour of existing practices in applying SWM, we move on to interrogate the concept in a broader context, exploring areas for its further application but also stepping back to look at present and potential limits to its application.

For starters, we posted on the IWRA website as our Question of the Year:

Is Smart Water Management Really a Smart Idea?

- If so, in what ways? If not, why not?
- Or is it smart in some ways and not in others?
- What could make it smarter (institutions, policy, other)? (www.iwra.org/questionoftheyear2019)

A sample of the 17 very thoughtful responses we received awaits the reader to launch the next part of the issue. To simplify grossly, the consensus answer is 'Yes, but …'.

We follow with a number of research articles and commentaries that look at SWM in three general ways: within the 'conventional' utility framework (Grigg, 2020); stretching the domain of SWM without questioning its purpose (Bjornlund et al., 2020; Moniz et al., 2020; Schweitzer et al.,2020); and stepping back and looking at SWM with a critical but not dismissive eye (Hartley & Kuecker, 2020; Trudeau, 2020).

Grigg (2020) sees SWM as definitely positive in its ability to increase the operational capacity of water utilities. He also sees great potential in an area where others have concerns about SWM – how it can improve access and affordability for end users, such as local communities that are at risk of being, or have already been, left behind. SWM could help those water providers with effective utility operations to leapfrog over the barriers to access that currently plague those communities.

Nonetheless, these anticipated long-term benefits are speculative. As in other areas where black box algorithms are coming to rule our lives, it will be necessary to address security and privacy issues.

Bjornlund et al. (2020) argue that SWM also has potential for small-scale irrigators in sub-Saharan Africa and, by ensuring consistent and ongoing use, can provide data for higher-level water planning and management. This presents additional challenges, as the technology needs to be used by often poor and illiterate farmers. In this context providing technology in isolation can be counterproductive, as it might not be sustainable and consistently adopted and used. To achieve this the technology must be people-centred, tailored in collaboration with end users (so it helps them achieve their objectives), easy to use, cost-effective and robust. The introduction of the technology must therefore use a two-pronged approach combining smart technology with innovative institutional processes that facilitate learning and capacity building, leading to behaviour change. Only if using the tools is beneficial to the end users will reliable data be collected on a large scale to inform higher-level decision making.

Schweitzer et al. (2020) describe how the United Nations High Commissioner for Refugees (UNHCR) has used SWM technologies to ensure delivery of safe drinking water to South Sudanese refugees in Uganda, where tanker delivery often did not reach the camps because drivers would sell some or all of the water in transit. UNHCR introduced technologies that can be used to monitor both water trucks and static water reservoirs, to ensure that water is delivered as intended and equitably distributed. The system is based on an open source software platform that has the capacity to visualize the data. UNHCR hopes to expand the system to include additional SWM sensor applications for measuring water quality, groundwater resources, and the operational performance of distribution networks.

Moniz et al. (2020) discuss a pilot project to use free and open access source software to improve the quality of data for urban water use on the Canary Islands. By digitizing the errors, this system can quickly identify issues that need to be addressed. The data can be integrated into other existing systems, and predictive tools can help in developing consumption plans and shared solutions. The authors argue that the pilot project has shown that the system enables flexible and reliable systems for advanced, tailored solutions to improve water quality and supply, save the utilities money, and improve the quality of service from the utility and its ability to inspect water infrastructure. It might also facilitate further automation of current manual operations.

In her contribution, Trudeau, 2020 argues that SWM is not a means of fixing systemic inefficiencies inherent in the traditional design of urban water infrastructure. SWM can support the existing infrastructure, but systemic inefficiencies need to be addressed as part of a long-term plan to reconsider the underlying approach to potable, wastewater and stormwater system designs. In many places, especially Europe and North America, water infrastructure was built in a way that does not recognize the cyclical nature of water. For instance, we need to rethink the assumption that wastes should be disposed to waterbodies, and that we should use potable water for purposes such as suppressing fires and flushing toilets.

This leads us in turn outside the confines of the water sector, to core issues of urban design and land use and their interconnection with waterways. We need to consider how we would design water infrastructure and support systems if we could do so from scratch, through developing a shared long-term vision involving political leaders, communities and experts, backed by appropriate funding commitments beyond the myopia of a short-term political or budgetary cycles. At the same time, existing infrastructure must be maintained, and solutions have to be tailored to individual watersheds and local water resources. This is a tall order – reimagining the city as well as its water – but unless issues such as these are addressed, SWM will continue to be applied to fundamentally inefficient infrastructure.

Similarly, Hartley and Kuecker (2020) explore the moral hazards of SWM as a technological fix for a broken system that does not address the cause of the problem, but allows society more time to continue convenient but wasteful habits. Measures of success are currently set by the SWM technologies. The authors argue that both hard (technological) and soft (behavioural and conceptual) approaches are needed to address problems. If critical thinking is not used to challenge and curb the technocratic powers of SWM there is a risk of social and environmental costs. Policy makers need to think of all possible narratives, consider what could go wrong, and assume that it probably will. Hence the precautionary principle should be applied to the planning, design and implementation of SWM solutions. There could be long-term consequences if we do not think outside the current lenses of goals (even the Sustainable Development Goals) and technologies.

We close with a checklist for you to review before you go (Kuisma, Nickum, et al., 2020). Then we send you on your way to do what you can to make a smarter water future.

Overall, this special issue provides a cutting-edge overview of existing practice, potential benefits, and the pitfalls of limiting our consideration of SWM to the current relatively narrow, technological definition. It makes some critical observations of measures that policy makers, planner and practitioners need to take to increase the likelihood of reaping the benefits while avoiding the pitfalls.

Disclosure statement

No potential conflict of interest was reported by the authors.

ORCID

Henning Bjornlund http://orcid.org/0000-0003-3341-5635

References

Bjornlund, H., Nickum, J. E., & Stephan, R. M. (2018). IWRA special issue: Wicked problems of water quality governance. *Water International*, *43*(3), 323–326. https://doi.org/10.1080/02508060.2018.1452864

Bjornlund, H., van Rooyen, A., Pittock, J., Parry, K., Moyo, M., Mdemu, M., & de Sousa, W. (2020). Institutional innovation and smart water management technologies in small-scale irrigation schemes in southern Africa. *Water International*, *45*(6). https://doi.org/10.1080/02508060.2020.1804715

Grigg, N. (2020). Smart Water Management: Can it improve accessibility and affordability of water for everyone? *Water International*, *45*(6). https://doi.org/10.1080/02508060.2020.1768738

Hartley, K., & Kuecker, G. (2020). The moral hazards of smart water management. *Water International*, *45*(6). https://doi.org/10.1080/02508060.2020.1805579

Kuisma, S., Clench, C., Garcia Quesada, M., Nickum, J. E., & Bjornlund, H. (2020). The report that sparked this special issue. *Water International*, *45*(6). https://doi.org/10.1080/02508060.2020.1826662

Kuisma, S., Nickum, J. E., Bjornlund, H., & Stephan, R. M. (2020). Before you go: The editors' checklist of what we now know about SWM. *Water International*, *45*(6). https://doi.org/10.1080/02508060.2020.1830580

Moniz, P. F., Almeida, J. S., Pino, A. T., & Suárez Rivero, J. P. (2020). GIS-based solution for urban water management. *Water International*, *45*(6). https://doi.org/10.1080/02508060.2020.1765130

Nickum, J. E., Bjornlund, H., Stephan, R. M., & Kuisma, S. (2020). Is Smart Water Management really smart? What experts tell us. *Water International*, *45*(6). https://doi.org/10.1080/02508060.2020.1826670

Schweitzer, R., Harvey, B., & Burt, M. (2020). Using innovative smart water management technologies to monitor water provision to refugees. *Water International*, *45*(6). https://doi.org/10.1080/02508060.2020.1786309

Tabuchi, J. P., Blanchet, B., & Rocher, V. (2020). Integrated smart water management of the sanitation system of the greater Paris region. *Water International*, *45*(6). https://doi.org/10.1080/02508060.2020.1830584

Trudeau, M. P. (2020). SWM and urban water: Smart management for an absurd system? *Water International*, *45*(6). https://doi.org/10.1080/02508060.2020.1783063

Villareal, F. G., Lartigue, C., Hidalgo, J., Hernández, B., & Espinosa, S. (2020). SWM technology for efficient water management in universities: The case of PUMAGUA, UNAM, Mexico City. *Water International*, *45*(6). https://doi.org/10.1080/02508060.2020.1830588

Yi, S., Ryu, M., Suh, J., Kim, S., Seo, S., & Kim, S. (2020). K-water's integrated water resources management system (K-HIT, K-water Hydro Intelligent Toolkit). *Water International*, *45*(6). https://doi.org/10.1080/02508060.2020.1830583

The IWRA report that sparked this book

Stephanie Kuisma, Callum Clench, Monica Garcia Quesada, James E. Nickum and Henning Bjornlund

To better understand how Smart Water Management (SWM) is being implemented around the world, and the challenges and solutions that are being addressed, the South Korean national water agency, K-Water, partnered with the International Water Resources Association (IWRA) to develop the *Smart Water Management Case Study Report* (https://www.iwra.org/wp-content/uploads/2018/11/SWM-report-final-web.pdf). The report provides a variety of SWM case studies from around the world, comparing their approaches, the technologies used, the lessons learned, and how their work helped address the UN's Sustainable Development Goals.

In the report, SWM solutions are presented from diverse projects across both developed and developing regions, and from the micro scale (e.g., households) through to the transboundary scale. These case studies demonstrate the potential of SWM solutions to assist in the advancement and integration of traditional water infrastructure in urban settings, while providing simple-to-use irrigation solutions in rural agricultural settings, along with support for improved planning, management and operation for water suppliers and users.

So what is smart water management?

SWM is the use of integrated, real-time information and communication technology solutions - such as sensors, monitors, geographic information systems and satellite mapping, and other data-sharing tools - in water management. Over the past decade SWM has become an area of increasing interest as governments, industries and utilities have moved towards real-time-data collection and use to optimize their operations and knowledge. SWM can provide them with integrated water management solutions, at all scales and across various contexts, to resolve water challenges in both developed and developing counties. The report demonstrates the numerous benefits reported across these case studies through the implementation of SWM, as shown in Table 1.

The potential application of smart systems in water management is wide and includes solutions for water quality and quantity, efficient irrigation, pressure and flow management, ecosystem protection, flood and drought management, stormwater and sewage management, planning, and much more. SWM can also address water infrastructure by integrating it into broader networks in order to share data to reduce water and energy

Table 1. Benefits of SWM implementation.

Social benefits
• **Access to clean water and sanitation** through water treatment and monitoring
• **Health improvements** through increased access to clean, safe water
• **Improved livelihoods** through job creation, greater opportunity for further education, higher productivity and other opportunities
• **Increased training and capacity building** for the local community and staff
• **Increased sharing of solutions** to support sustainable development
• **Increased decision-making opportunities** through increased engagement and knowledge sharing
• **Greater collaboration with communities** through engaging with local stakeholders at the beginning of the project
• **Greater security** by improving water security and resilience to climate change
• **Increased trust** in water suppliers and the safety of water sources
• **Improved access to data and information** through real-time data sharing with all water users
• **Increased gender equity** through increased opportunities for capacity building and further education
• **Reduced conflict over water access** leading to more trust and more willingness to engage in collective action.
Economic benefits
• **Increased efficiency** in irrigation systems and wastewater treatment systems
• **Reduced waste** through the reduction of water loss through leakage
• **Job and opportunity growth** through job creation for SWM project research, design, development and implementation
• **Improved capacity** of water systems to manage flows and reduce damage during storms and floods
• **Reduction in future infrastructure costs** by integrating smart technology tools to improve capacity and efficiency, reducing the need for additional infrastructure
• **Mobilization of funds** from public and private sources, as well as international funding sources.
Environmental benefits
• **Improved water quality** through reduced pollution and contamination in waterways
• **Improved ecosystem health and protection** through improved water quality and quantity
• **Reduction in groundwater depletion** through reduced over-abstraction
• **Reduced land degradation** through flood and drought management and reduced nutrient loss in the soil
• **Reductions in CO_2 emissions** through energy optimization and reduced energy consumption
• **Reduced water consumption** through leak detection, reduced demand and increased reuse.
Governance benefits
• **Improved management and knowledge**, as measurement is critical for effective management
• **Improved accuracy of data**, as real-time data should also be SMART (specific, measurable, actionable, relevant, and time-bound) data
• **Increased community-led decision-making opportunities,** as water users can make decisions based on real-time water use and quality information
• **Improved transparency,** as water users have real-time access to water use and quality informaiton.
Technology benefits
• **Opportunity to test and develop** new and innovative tools for water management
• **Creation of innovative technologies**with the potential for commercialization
• **Identification of the remaining gaps** in technology adoption (e.g., standardization of software and tools to make it easier to adopt the right mix of tools for each situation)
• **Showing the potential for SWM tools** to deliver successful outcomes and in turn significant social, environmental, governance and financial impacts.

consumption, provide targeted irrigation for agriculture, and make wastewater treatment more efficient. In developing countries, it is specifically relevant for taking into consideration urban and regional data and water consumption so that governments can improve public health through the possibility of checking water quality, water resource availability and water distribution in all the neighbourhoods of developing cities, towns and rural communities, especially where informal dwellers are settling.

SWM is a high-tech strategy to deal with economic, social and environmental urban and regional issues as a way to better use our water resources while protecting the most vulnerable places and creating innovative types of economy and management. At a time when data are part of everyday life, it is a natural step for decision makers to include SWM in their policy strategies in order to provide a more adapted response to urban and regional organization. Taking into consideration the current issues communities face and the global commitments under the UN's Sustainable Development Goals, SWM can provide support for any contemporary Integrated Water Resources Management strategy. Thus policy makers around the world are increasingly integrating smart principles into their urban, rural, regional and national strategies, which will in time provide better understanding of the dynamics of cities and regions, leading to more resilient, sustainable and safer living environments. The key factors for the successful implementation of SWM as seen throughout the case studies within the report are shown in Table 2.

While interest in smart management has increased rapidly over the past decade, its adoption in water policy has been slower than in other sectors, such as energy and transport. To correct this, decision makers must have information on the benefits of SWM and how policy can support successful SWM implementation. It is therefore essential that reports such as the present one provide these insights for policy makers, while also sharing the knowledge with the broader water community interested in implementing SWM solutions (see Table 3).

Table 2. Factors for success.

Cross-cutting factors

- Political commitment from government at all levels
- Support from national government policy, legislation and regulation
- Two-pronged approach (combining the use of SWM tools with engagement, governance and/or a strong business model) to support the implementation and increase the adoption of, and positive outcomes from, SWM technologies
- Strong stakeholder engagement from the beginning of the project across and within sectors, especially including local agencies and communities, to ensure active community participation and decision making
- Multidisciplinary approach (both across sectors and within sectors) to ensure all factors can be taken into account (e.g., environmental, technical, scientific, policy, regulation, financial, maintenance).

Social factors

- Active stakeholder engagement from the beginning of the project
- Local stakeholders to be involved in decision making and implementation
- Improved livelihoods from job creation and increased opportunities such as time for further education and skill development
- Increased community trust of water suppliers and water resources
- Education, training and capacity building for local communities.

Economic factors

- Long-term investment to enable ongoing research, development, testing and implementation to support taking SWM solutions to market
- External financial support from both public and private investors to assist in project implementationin the short term
- Consideration of the non-financial benefits (e.g., environmental, social, governance), which are often apparent in the short term, alongside the financial returns, which are medium-to-long-term
- Strong business cases to support replication and scaling
- Demand management and improved efficiency as a means to water and energy savings.

Environmental factors

- Regulations, economic instruments and information to encourage behavioural changes to improve water quality, water use efficiency, and natural resource protection
- National plans to improve/resolve water challenges
- Commitments from international funding bodies to meet and address the Sustainable Development Goals, including water
- Commitment from leading organizations and stakeholders to address these environmental challenges.

Technical factors

- Allowing adequate time to design, develop, test and adjust technology for better results
- Undertaking a baseline assessment of the challenges and what needs to be addressed to ensure that the right mix of technology and non-technological solutions are implemented
- Collaborating with all sectors to ensure adequate and accurate data (e.g., electricity data) are shared to support decision making
- Integrating smart tools and systems across networks to enable collaborative decision making
- Integrating smart tools with traditional infrastructure
- Willingness of water utilities and governments to test the possibilities of smart technologies.

Table 3. Policy recommendations.

Strategy	Policy directions
Society: SWM for better quality of life	(1) Facilitate adoption of SWM tools, especially in developing countries, to support access to basic services and to support equality for poverty reduction, public health and quality of life. Include capacity development, technology sharing, collaborative business models and community governance and decision-making opportunities. (2) Build trust and community engagement using SWM tools in areas where people feel unsafe using the local water sources. (3) Empower people in developing countries with smart tools to reduce the time spent on water management and increase farm income and time available for other activities (e.g., further schooling or additional work opportunities).
Economy: Investment in SWM for improved resilience and sustainable development	(4) Strengthen collaboration across and within sectors to provide opportunities for networks to share information and data for effective and efficient water management. (5) Value non-financial benefits (e.g., environmental, social, governance and technical benefits) as equally important as financial benefits for SWM implementation, as they contribute to building resilience to the effects of climate change and growing populations. (6) Support long-term investments in SWM implementation to enable adequate research, development and testing.
Environment: SWM for protecting and conserving water resources and ecosystems	(7) Introduce policies, regulations and incentives to drive environmental and ecosystem protection through use of SWM. (8) Encourage SWM solutions to improve water quality, manage demand and use, support water reuse, reduce groundwater depletion, increase energy efficiency, etc. (9) Introduce SWM solutions for climate adaptation plans for flood and drought planning and management and major storm events.
Technology: To support evolving smart technology development and adoption	(10) Develop standards to ensure all SWM technologies are compatible (can communicate) with each other to enable tools to be purchased across various suppliers and enable those implementing SWM to create the right set of tools for each context. (11) Support ongoing research, testing and development of SWM tools to advance them to point where they are robust, require minimal maintenance, and are ready to be commercialized (government policies that support taking SWM tools from R&D to market). (12) Support technology to assist in regions without built infrastructure of the adequate resources (e.g., electricity), as currently SWM infrastructure is almost always dependent on built infrastructure.
Governance: Building network capacity for increased resilience and collaboration	(13) Empower people, especially those in developing countries, by providing them with SWM tools, data and capacity development and education to enhance/support local decision making. (14) Strengthen the capacity to adapt to climate change by adopting SWM planning and operational technology. (15) Plan for water disasters in advance, creating proactive policies instead of reactive policies.

The sharing of SWM successes (and challenges) also helps water industries, utilities and other users better understand the enabling facts and the barriers to successful SWM implementation, leading to a greater uptake of successful SWM projects in the future. As SWM has continued to grow over the past decade we have seen a variety of SWM solutions across all scales for a wide range of water challenges and contexts. To foster the continued growth of SWM and to promote innovation and smart solutions for future water management, the report aims to share these insights from around the world.

Summary of findings from the report

SWM technology by itself will not always resolve the water challenges faced by a project. In some cases, a two-pronged approach is necessary to address the complex nature of each challenge. The second prong can include community engagement, governance schemes or business models and is as important to the success of many projects as the SWM tools themselves.

The projects included in the report show the considerable potential for SWM to address numerous water challenges, across various scales, geographic locations and developing and developed regions, while also offering social, economic, environmental and governance benefits. These projects have also demonstrated the enormous potential of SWM to assist with achieving the SDGs, across a number of the goals and targets.

While it is important to recognize that each project is set within its own context, the overarching lessons that emerge in the report highlight the similarities between case studies to show how SWM can be successfully implemented around the world, along with the challenges that remain.

As SWM is still an emerging field, these projects demonstrate the untapped potential of what can be achieved using innovative SWM technology and solutions. As the field progresses and technologies evolve, the potential for SWM adoption across all contexts will continue to grow, leading to increased opportunities for both developed and developing regions, and innovative solutions for our current water challenges.

In order to continue learning from these case studies, it is important to follow them on their journey to see how challenges are addressed as the technology evolves, and what impact introducing SWM continues to have in their region. This is important when trying to scale up or down or to transfer existing SWM solutions to new locations, bearing in mind the necessary adaptions to the local context and challenges.

It will also be interesting to see how SWM technology and solutions can move from the research and development stage to the testing stage and finally to market – in other words, how SWM can become self-sustaining without depending on government support in the early phases.

At this stage, many of these projects have shown the potential for SWM technology to successfully resolve water challenges. It is important now to develop the businesses cases for adopting, scaling and transferring these solutions. This is why the monitoring and measuring of SWM benefits must continue. The next phase of research would be aimed at helping capital investors see the benefits and potential of SWM, opening possibilities for future investment.

Now that a wider variety of smart tools are on the market, integrated smart networks will start to emerge, and with them increasing opportunities for sustainable cities and regions to integrate their various smart infrastructure, such as smart energy grids. While retrofitting existing cities is possible, the opportunity offered by urbanization and the creation of new cities and suburbs means that these new urban environments offer the greatest potential for smart technology integration.

The report demonstrates how far SWM has already come in a short time and the considerable benefits it can provide in both developed and developing regions, especially when coupled with strong policy support and community engagement. It also explores some of the constraints and barriers encountered to date. In the end, however, it is certain that SWM has nearly unlimited potential to contribute to the realization of the goals of Integrated Water Resources Management and sustainable development through smarter management of water.

What next?

Following on from the successful delivery of the case study report, the IWRA, K-Water, and the Asian Water Council have agreed to initiate a new phase of collaboration on a three-year joint research project, Smart Water Cities. This project aims to identify and examine examples of smart water technologies in cities around the world, and to collect and analyze a set of global standards frameworks and certification schemes applicable to smart and sustainable cities. In this sense, this new project builds on the findings of the present report to focus on the specific opportunities and challenges of employing SWM technologies in new smart urban developments. The project, which started in July 2020, will connect researchers around the world who are working on the topic and produce three main reports focusing on Identifying Solutions, Measuring Impact, and Certification.

Disclosure statement

No potential conflict of interest was reported by the authors.

ORCID

Henning Bjornlund http://orcid.org/0000-0003-3341-5635

Reference

International Water Resources Association and K-Water. (2018). *Smart water management report.* Korea Water Resources Corporation (K-water). https://www.iwra.org/wp-content/uploads/2018/11/SWM-report-final-web.pdf

Part II

Case Study Summaries

SWM technology for efficient water management in universities: the case of PUMAGUA, UNAM, Mexico City

Fernando González Villarreal, Cecilia Lartigue, Josué Hidalgo, Berenice Hernández and Stephanie Espinosa

ABSTRACT

This case study details the establishment of a real-time water quality and quantity monitoring system at the National Autonomous University of Mexico (UNAM), Mexico's largest university based in Mexico City. It is structured to outline the water challenges facing Mexico City and UNAM, the solutions implemented by PUMAGUA, the Program for the Management, Use and Reuse of Water at UNAM, and the lessons learned in the process which can be extended beyond the university.

Summary

Mexico's largest university (and indeed the largest university in Latin America), the National Autonomous University of Mexico (UNAM) faces challenges of water quality, access and inefficient management reflective of Mexico as a whole. To address them, UNAM launched the Program for the Management, Use, and Reuse of Water (PUMAGUA) in 2008 with the aim of implement efficient management of water at the campuses of UNAM. The three key objectives of PUMAGUA were to: 1) reduce potable water consumption by 30% and water losses by 50% through improved practices and leak detection, 2) improve the quality of drinking water and treated wastewater in accordance with Mexican regulations and 3) promote participation of the entire UNAM community in the efficient use of water. Prior to PUMAGUA, UNAM faced up to 50% water loss through leakages in the university pipelines, had irregularities in drinking water disinfection, was out of compliance with regulations on treated wastewater, and lacked continuous capacity development activities or communication campaigns to save and conserve water.

In the decade since PUMAGUA was initiated in 2008, it has achieved a reduction of 25% in total water use in the main University City (CU) campus, despite a population increase of more than 37% from 2008 to 2018. Drinking water and treated wastewater are now both of excellent quality. PUMAGUA has also enhanced the responsible use of water by key actors, as well as the participation of students and lecturers in generating proposals to solve water problems. In addition, it has carried out work-shops targeted at maintenance staff and gardeners to enhance water saving actions.

In order to produce these results, PUMAGUA has made use of Smart Water Management (SWM) technology, in particular, remote water consumption measurement and water quality assessment tools. The Program created UNAM's Observatory of Water, a digital real-time platform that includes water quantity and quality data, and social participation, in order to respond promptly to any eventuality and to actively interact with the university community. PUMAGUA has extended its activities to trial smart water projects in another university campus and also in a low-income housing district in Mexico City.

Several lessons can be l earned from the PUMAGUA program, including 1) smart water management can build trust in water users (in our case, in the university community); 2) smart technology requires a lot of time and effort to manage and maintain, especially in the early stages of a project; and 3) given the speed and scope of technological change, it is essential to have the financial resources to acquire technology upgrades, as well as the human capabilities to manage those updates. In the future it is planned to update the SWM technology for water consumption measurement and use this kind of technology for other purposes, such as measuring soil moisture.

Background

In 2013, to encourage tap water consumption while reducing bottled water consumption, the Legislative Assembly of Mexico City modified the city's Law of Education to require the government to instal water fountains in primary and secondary schools (ALDF, 2013). In 2016, at the national level, the Ministry of Public Education modified the Law of Physical Education Infrastructure, making it an obligation for public schools to provide sufficient drinking water for students and to instal water fountains, according to guidelines established by the Ministry of Health in collaboration with the Ministry of Education (DOF, 2016).

The University City at UNAM

The National Autonomous University of Mexico (UNAM), in Mexico City (Box 1), is the largest university in Latin America (Figure 1), with a population of 185,000 people during the school term at the 700-hectare main campus alone. Its main purpose is to train professionals, to organize and carry out research focused on national conditions and problems and to disseminate the benefits of that research to the country (National Autonomous University of Mexico, 2017). After the Fourth World Water Forum, in Mexico City in 2006, the university created the UNAM Water Network, with the participation of 26 schools and institutes in the university. Yet while UNAM was conducting research on water quality challenges throughout Mexico, it did not have a management system in place for its own university premises. Hence in 2008 PUMAGUA was established, with an interdisciplinary team assembled to implement an integrated programme of water management, use and reuse on UNAM's campuses, with the participation of the community.

Box 1. Water challenges in Mexico City.

High population density causes intense pressure on water resources in the region of Mexico where Mexico City and the University City are located.

In 2008, per capita average tap water consumption of water in Mexico City was at 184 L per person per day, but consumption in different areas of the city ranges from less than 125 to over 475 L per person per day (Capella- Vizcaino et al., 2008). Yet while most households have taps, water utility networks in Mexico City, as in the rest of the country, are unreliable for many. González Villarreal et al. (2015) found that at least a quarter of households do not receive water daily, and Capella (2015) found that in Mexican cities 30–50% of water is lost to leaks.

Partly in consequence, Mexicans have become the leading per capita consumers of bottled water in the world, each using 215–234 L/y on average. Approximately 80% consume bottled water, at considerable cost, instead of inexpensive tap water. Low-income households spend even more on bottled water because their water access is inadequate, due in part to rationing of water by the utilities (Torregrosa, 2012).

Mexicans buy bottled water mostly because they believe that the tap water is not safe (González Villarreal et al., 2015). A study in Mexico City by the Inter-American Development Bank found that 81% of interviewees buy bottled water because they distrust tap water, even though 88% of them stated they had never fallen ill after drinking it.

Romero Sánchez (2015) has shown that 12% of inhabitants (approximately 1 million people) do not receive good- quality water. According to Romero-Lankao (2010), even when water quality is allegedly monitored on a continuous basis, it is doubtful that this monitoring is effective, particularly in central areas of the lake system and in some areas of its aquifers. Those aquifers are polluted with bacteria, faecal matter, sulphates and other things due to over-exploitation, subsidence, fractures, lack of access to sanitation for some sectors, and lack of maintenance of domestic installations, such as water tanks. There are no accurate data on the health implications of the poor-water quality in some areas of the Metropolitan Area of Mexico City, but diarrhoeal diseases are the fourth most common cause of child mortality in the city.

Figure 1. Location of the University City (Ciudad Universitaria, CU) in Mexico City.
Source: PUMAGUA (2012).

Water supply in UNAM

The Directorate of Works at UNAM is responsible for operating the water services on UNAM's campuses, with the water supplied from three wells on the campus. CONAGUA has granted a concession over the wells to UNAM, which is not required to pay for the supply and use of water. This is of particular importance because, unlike in many other SWM projects, reducing water consumption did not reduce the agency's costs.

Water conditions at the CU when PUMAGUA was launched

Soon after PUMAGUA was launched, the team realized that, in addition to calculating the water balance, it was fundamental to assess the quality of drinking water and treated wastewater in order to determine compliance with Mexican regulations and to improve water safety. In addition, the team understood that technical improvements both in water quality and in water saving would need community support in order to be long-lasting. Therefore PUMAGUA was organized into three main areas: water balance, water quality, and communication and participation, each addressing a key objective:

- To reduce tap water consumption by 50% through improved practices and leak detection;
- To improve the quality of drinking water and treated wastewater in accordance with Mexican regulations; and
- To promote participation of the entire UNAM community in the efficient use of water.

It was intended that PUMAGUA would become a pilot case model of water management that could be exported to other UNAM campuses, to other universities, and to other localities in the country. UNAM has six campuses throughout the country, 17 schools in the Metropolitan Area in Mexico City, and research institutes and schools in 20 states of Mexico, the United States, Canada, Spain and China. In addition to the pilot case, some SWM techniques were also implemented on other UNAM campuses.

The water challenge

The first activity of PUMAGUA, in 2008, was a diagnosis of the water system at the CU focusing on three core issues: water quantity, water quality and social participation. This was the first time such a diagnosis had been carried out.

A summary of the findings is as follows.

- There was no updated hydro-sanitary information.
- There was no water consumption measurement in buildings.
- The water losses in the distribution network were over 50%.
- There was non-compliance with the Mexican standards for quality of drinking water and treated wastewater.
- Three treatment plants were in need of refurbishment.
- The 26 decentralized anaerobic bioreactor water treatment plants, one at each college, needed to be replaced by a centralized system to guarantee water quality and optimal performance.

- There was no programme to elicit participation of university schools, institutes and offices in responsible water management and use.
- The CU community lacked knowledge about the water management system.
- Academics were reluctant to adopt water conservation measures that would involve some level of compromise.
- Students exhibited the highest willingness to participate.
- The system did not promote any specific irrigation or water-saving method.

Water quantity

To assess the volume of water used on site, the Directorate of Works analyzed the information available on the water distribution and wastewater collection networks, and measured drinking water supply and leaks using water meters.

A lack of hydro-sanitary information was immediately detected, including the layout of the distribution network. The only blueprints available were on paper, as there was no online system available to upload them to (Figure 2). Consequently surveys were made to determine the pipeline routes and to draft and digitalize hydro-sanitary blueprints. A total of 54 km of drinking water pipelines were identified. It was also found that there was no water consumption measurement within the buildings; none of the water meters installed in the past were working.

Thirty percent of the toilets and faucets (500 out of 6000) were not working properly or were not complying with the standards of Mexican authorities. After the assessment, a full inventory of infrastructure was created. The water system identified is briefly described in Box 2.

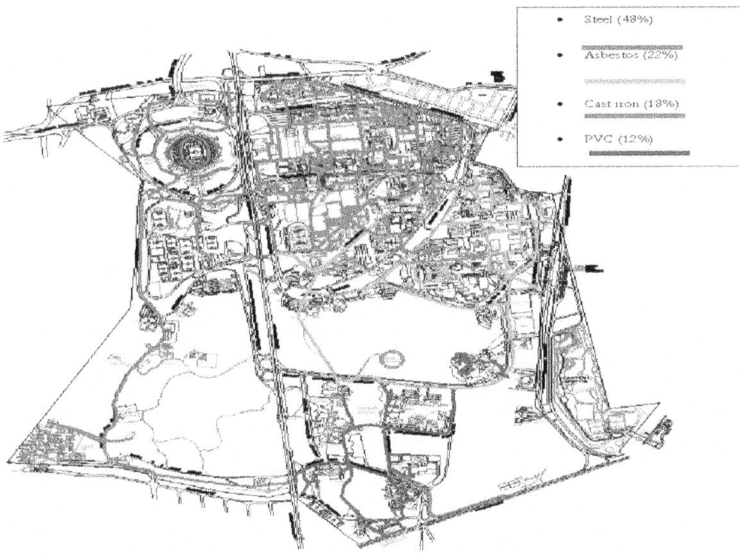

Figure 2. Digitalized water supply network of the University City. Pipeline colours correspond to different materials: steel, asbestos, cast iron, and PVC.
Source: PUMAGUA (2012).

Box 2. Water infrastructure prior to SWM implementation.

Drinking water infrastructure:

- 3 wells, with average extraction of 100 L/s
- 3 storage tanks with a capacity of 12,000 m^3
- 54 km of pipelines.

Wastewater infrastructure:

- 18 km of sewage pipelines
- 1.8 km of rainwater drainage
- 3 treatment plants and 26 small, integrated anaerobic bioreactor plants
- 13 cisterns of treated wastewater for irrigation of the university gardens.

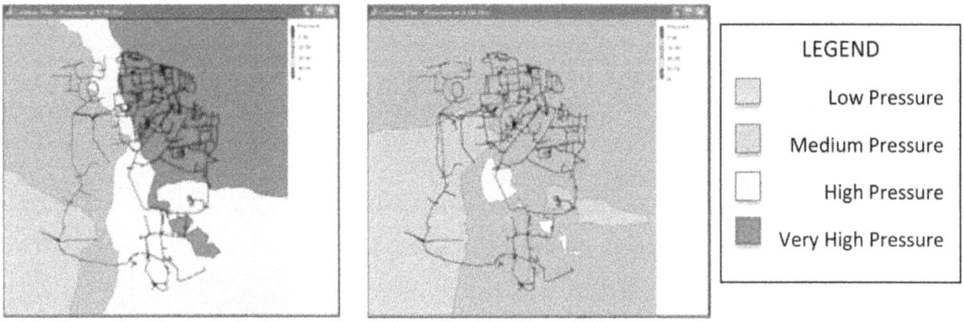

a) b)

Figure 3. Simulation of the distribution network of the University City of UNAM, (a) without pressure control, and (b) with pressure control (PUMAGUA, 2012).

Pressure modelling and sectorization

Pressures in the distribution network were modelled using EPANET software, and five hydraulic sectors were defined to facilitate leak detection (Figure 3). According to measurements by PUMAGUA, over 50% of the water extracted from the wells was lost through leaks and inadequate operations (e.g., 5000 litres of water per day were lost by laboratories that did not have recirculation systems for wastewater). Figure 4 shows the water balance in 2008.

Water quality

Water quality at CU was analyzed both before and after treatment, to determine whether there were pollution threats in the groundwater and whether the disinfection system worked properly. Periodic sampling of groundwater was performed from 2008 to 2010.

This allowed the identification of potential health hazards, such as total coliforms and faecal coliforms. Nitrate levels were close to the limits established in national regulations (NOM-127-SSA1-2000: 10 mg/L nitrates).

Water quality after treatment with chlorine was also determined. Samples were sent to an external certified laboratory for analysis of the 41 parameters included in the Mexican drinking water standard. All parameters complied with the standard, except for free residual chlorine, which is necessary to prevent the growth of microbiological organisms.

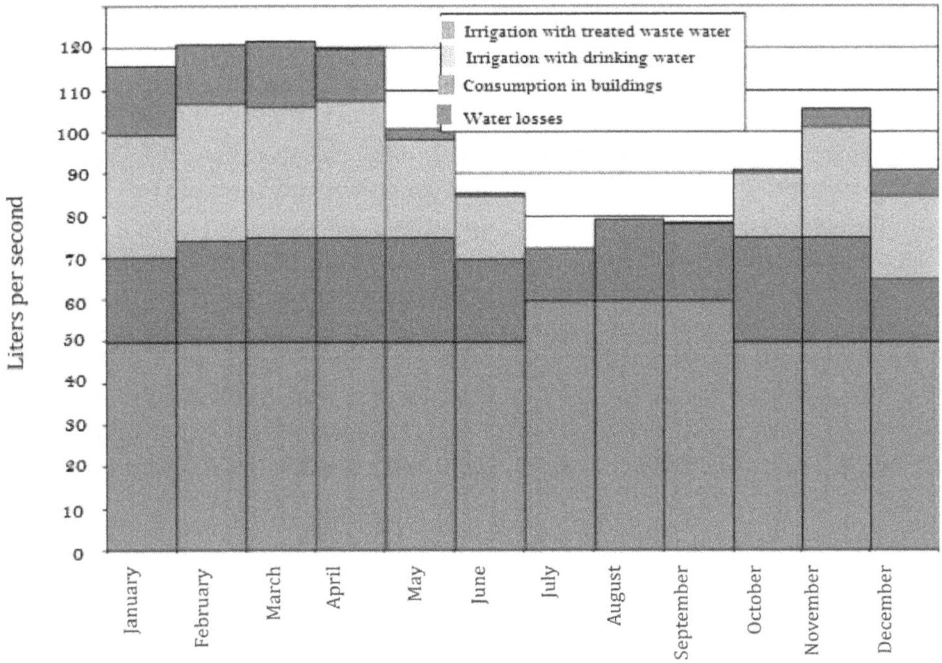

Figure 4. Water uses of University City, in 2008, showing water losses over 50%.
Source: PUMAGUA (2008).

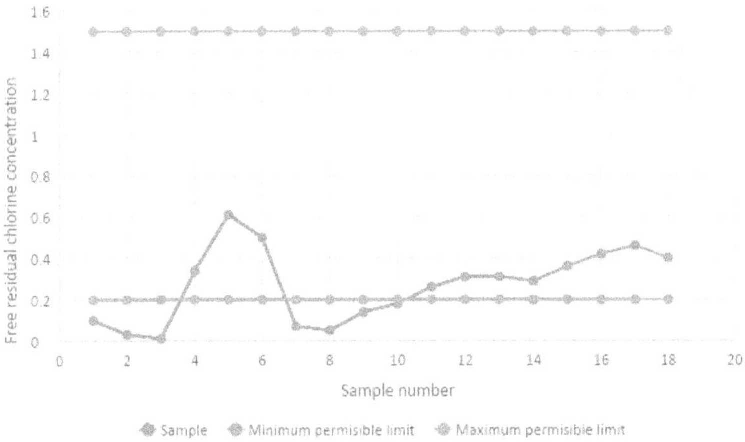

Figure 5. Concentration of free residual chlorine in 18 samples of disinfected water at Universum Museum, University City, UNAM, in 2008.
Source: PUMAGUA (2008).

It was below the minimum permissible in some samples. Figure 5 shows the free residual chlorine concentration in one of UNAM's buildings.

To assess the water quality on site, an *in situ* water sampling programme was also carried out, assessing the water quality in the water wells, the distribution system and the drinking water tanks (Orta de Velásquez et al., 2013). The sampling points were selected based on

Mexican standards, empirical criteria and the application of probabilistic systematic sampling. Initially 20 sampling sites were established to cover the five hydraulic regions, and testing was performed monthly. Concentrations of faecal coliforms, total coliforms, and free residual chlorine were assessed.

The main treatment plants and their effluents were also evaluated. The Directorate of Works operates three wastewater treatment plants around the campus. The team determined that water quality from the three plants did not comply with Mexican regulations and recommended refurbishment of the three plants (PUMAGUA, 2008).

In 2008, there were also 26 small integrated anaerobic bioreactor plants spread around the campus. Due to lack of maintenance, their water quality also did not comply with Mexican regulations, and PUMAGUA suggested closing all of them.

The distribution system for treated wastewater was also analyzed. In the campus water distribution system, treated water is pumped to cisterns, from where it is distributed for the irrigation of gardens. Every month since 2008, the effluent's water has been analyzed for faecal coliforms, helminth eggs, fats, oils and greases, biochemical oxygen demand, chemical oxygen demand, and total suspended solids. The PUMAGUA team assessed the physical and operational conditions of the plants and concluded that they were operating below design capacity, they were in poor physical condition, and their effluent quality did not comply with Mexican regulations (Orta de Velásquez et al., 2013).

Likewise, inspections of the storage cisterns revealed that most of them were in poor physical condition (lack of lid, leaks, rusted metal parts, etc.). Not surprisingly, the water stored in the cisterns did not comply with local regulations (Orta de Velásquez et al., 2013).

Social participation

PUMAGUA's initial diagnosis of the water system at UNAM CU also revealed that there was no programme within the University to encourage authorities of schools, institutes or administrative offices to implement actions to reduce water consumption or water pollution in school buildings.

A stratified survey was developed in 2009 by the PUMAGUA team to establish a baseline of community members who participate in responsible water management and use. The variables assessed were knowledge, beliefs, attitudes, practices, and preferred means of communication. The team believed that communication and training activities could be better designed if these variables were determined.

Face-to-face interviews were carried out, applying a questionnaire to 1480 people of different sectors (academics, students, administrative workers, laboratory staff, housing residents, and visitors), and special workshops took place for gardeners to determine their watering methods.

The main findings of these activities were:

- There was limited knowledge of the water management system in all sectors.
- Less than half of the interviewees noticed water wasting on campus.
- A significant portion of the participants (over 40%, particularly administrative staff) believed that UNAM's authorities did not care about water wasting.
- Students were the only sector that specifically recognized their own responsibility in water wasting.

- Academics were the sector with the least willingness to accept measures for water conservation that would require some level of compromise.
- Only a third of academics had been trained to adequately dispose of chemical residues in laboratories.
- Very few maintenance workers had received any sort of training on responsible water management.
- Multiple watering methods were used, without detectable criteria and without the purpose of saving water. Each gardener did as they thought best, in terms of frequency and duration of watering, to keep the gardens green.
- Bottled water use at UNAM was high, with 80% of those surveyed drinking bottled water instead of tap water (Box 3).

These results display the low engagement of the UNAM population with the importance of water conservation.

Smart Water Management solutions

The smart solutions integrated into PUMAGUA fall into three main areas: water quality, water quantity, and community participation (Table 1).

Water quality solutions

Operation of the real-time water quality monitoring system

For reliable and constant information about water quality, a real-time monitoring system was installed in the Institute of Engineering. Real-time monitoring was implemented by means of a data collection sensor system in one of the buildings of the Institute of Engineering of UNAM. For economic reasons, only one of these systems could be installed. The sensors measure six physicochemical parameters (free residual chlorine, conductivity, nitrates, pH, turbidity and temperature) through a constant water flow from the distribution network. The system collects measurements of these parameters

Box 3. Bottled water versus tap water consumption in CU, UNAM, in 2012.

In 2012, according to Espinosa-García et al. (2015), only 14% of the university community consumed tap water exclusively, while 75% consumed bottled water, and 11% consumed both. Bottled water was preferred mainly for organoleptic reasons (taste, colour, turbidity, odour) – 54% of respondents cited this reason, while health was the second reason (26% of respondents).

Table 1. Smart Water Management solutions.

	Water problem addressed		
SWM tool	Water quality	Water quantity	Participatory community
Water Observatory	X	X	X
Real-time monitoring sensors	X	X	
Telemetry system		X	
Outreach activities			X

every five minutes and sends them through a local area network to a website provided by the manufacturer.

The sensor system consists of a controller and four sensors, one for each physico-chemical parameter, with the exception of pH (measured by the chlorine sensor) and temperature (measured by the conductivity sensor). Water from the distribution network is fed to each sensor from a tap connected to PVC pipes (Figure 6, Box 4).

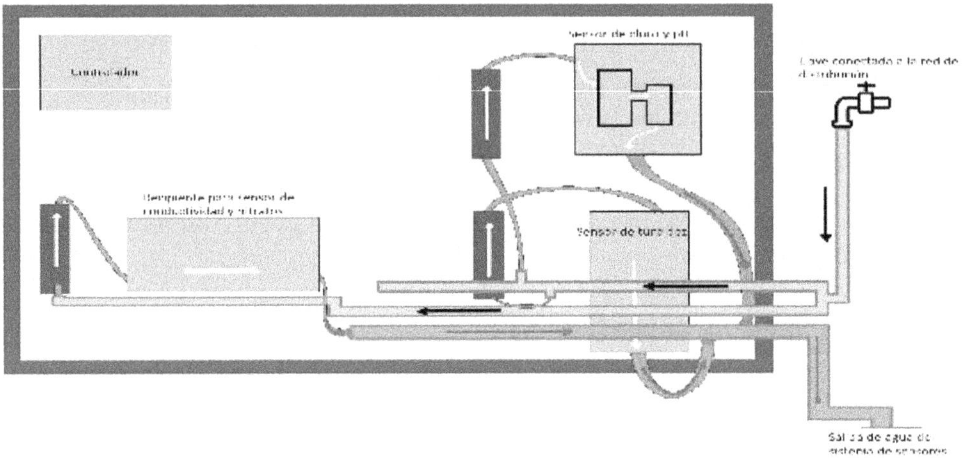

Figure 6. Water inflows and outflows in the real-time water quality monitoring system. Water from the distribution network feeds this system. Black arrows represent water flowing into a sensor, while red arrows represent water flowing out of the system.

Box 4. Sensors.

Conductivity and temperature

This sensor generates a low current whose intensity is then measured. The current corresponds to the conductivity of the volume of water in which the sensor is submerged (Figure 7). This sensor also has a resistive temperature detector.

a b

Figure 7. The conductivity sensor of the real-time water quality monitoring system: (a) the sensor; (b) a container built to allow constant water flow.

Nitrates
The nitrate sensor photometrically measures the concentration of nitrates in the water. Like the conductivity sensor, the nitrate sensor must always be submerged in water (Figure 7(b)) so that it can use ultraviolet light. Two receivers then measure absorbance: one of them works as a control element, while the other functions as a measurement element. The measured absorbance corresponds to the difference between the emitted light and the received light after its interaction with the disolved nitrates.

Free residual chlorine and pH
The free residual chlorine and pH sensor measures the concentration of free residual chlorine in water. This compound is made of dissolved chlorine (in low pH conditions), hypochloric acid gas and hypochlorite ions. The concentration of each of these elements depends on pH and temperature. As the concentration of hypochlorite ions depends on pH, this sensor has a potentiometer. The sensor comprises a gold electrode, a counter electrode made of silver, an electrolyte (potassium chloride solution), and a micropore membrane, which is selective to hypochloric acid (HOCl). When the water sample is in contact with the membrane, HOCl molecules diffuse into a narrow area between the membrane and the cathode, which contains potassium chloride. At this point, a constant potential is applied to the cathode, and the HOCl consequently decomposes into chlorine ions and water. The system calculates free residual chlorine using the dissociation curves of HOCl.

Turbidity
The turbidity sensor measures water turbidity by emitting a beam of light that passes through the water sample and using a photocell to detect the light reflected by the particles at a 90° angle from the emitted light. The greater the amount of suspended particles, the greater the amount of light reflected towards the photocell and the higher the turbidity of the sample.

Water quality in UNAM's online Water Observatory

The Water Observatory is an online platform that hosts the real-time data for the water users at UNAM. It has two modes: administrator (accessible only by the staff at PUMAGUA) and user (accessible to anyone with access to the internet). The two modes were designed to enable PUMAGUA to store all of the data in one location, while providing relevant information to the community. The decision to provide different levels of access for the two modes was made as a result of the abundant volume of data provided to the platform, and the need to allow the regular user to quickly access information. Also, much of the technical information may be difficult for the general public to understand and might cause unnecessary alarm, while it is of the utmost importance for PUMAGUA to be aware of these data, in order to take action in case of eventualities.

The Water Observatory summarizes the results of the water quality analysis. Drinking water results are displayed based on three types of monitoring: real-time information from sensors, on-site sampling by PUMAGUA, and on-site sampling by an external certified laboratory. It also hosts results from the on-site sampling of treated wastewater in the Cerro del Agua treatment plant and six wastewater storage tanks.

On the left side of the main screen, a map of the main campus of UNAM is displayed, in which the items of interest are depicted. Information about the date of monitoring, type of water, and the water source is linked to each element displayed in the map. The right side of the map provides an option menu. In administrator mode, it offers 10 search criteria for specific consultations for each site:

(1) Campus (currently the information comes from the main campus only, but in the future other campuses can be included)
(2) Type of water: drinking water or treated wastewater

(3) Use of water: human use and consumption (for drinking water only), or reuse in gardens (for treated wastewater only) – in future, reuse in toilets may be added
(4) Applicable regulation: drinking water quality, treated wastewater quality, or conditions of storage tanks
(5) Type of monitoring: real time or *in situ* sampling
(6) Water source: specific well, for drinking water; or specific treatment plant, for treated wastewater
(7) Monitored element: drinking fountain, distribution network, intake, storage tank, or regulation tank
(8) Location of monitored element: school, institute, or administrative office
(9) Parameter: includes the five parameters measured in real time, the three parameters monitored *in situ* by PUMAGUA, and the 41 parameters analyzed by the external laboratory
(10) Period.

For the real-time sensors, the platform can show daily information, as well as data from several days, months or years. A graph can also be displayed.

In the user mode, only five options are shown, all related to drinking water:

(1) Campus
(2) Type of water: drinking water
(3) Type of monitoring: real time or *in situ* sampling
(4) Monitored element: drinking fountain or distribution network
(5) Location of monitored element.

The site map is useful, as it shows the real-time water quality of each element (e.g., drinking fountain, intake, or distribution network) on the campus. When the water quality of an element complies with the Mexican national water regulations, it is depicted in green; if not, it is depicted in yellow. A pie graph showing the percentage of total data complying with Mexican water quality regulations is shown in Figure 15. This screen shows the drinking devices whose water complies with regulations, i.e., tap water (green dots), and whose water is not to be consumed until advised due to bacterial contamination (yellow dots). The pie chart shows the percentage of compliance (100% green).

Statistical analysis can be performed with all the information stored in the database since 2013. Correlations and variance analysis can also be carried out using the real-time data.

Addressing water quantity

Water consumption measurement points

Measurement points to assess water consumption were selected according to criteria established in various handbooks, such as the (Mexican) *Handbook of Drinking Water and Sanitation* (MAPAS for its acronym in Spanish) and the handbook of the National Water Commission. Electromagnetic water meters were used for macro measurement (wells, tanks, hydraulic sectors), while for water intakes of buildings (micro measure- ment) volumetric meters were installed (Figure 8).

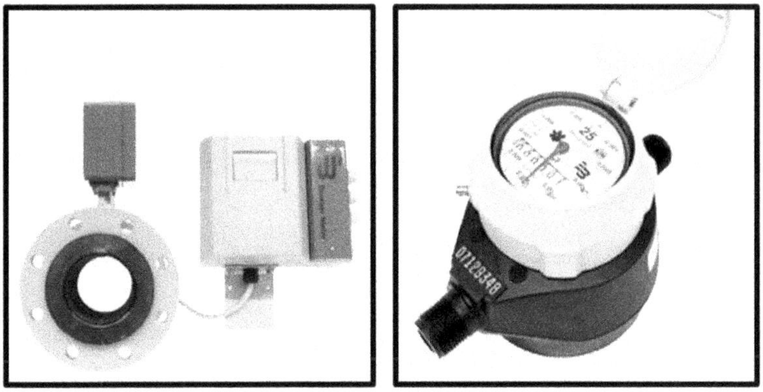

Figure 8. Electromagnetic water meter (*left*) and volumetric meter (*right*).
Source: PUMAGUA (2012).

PUMAGUA has installed nine macro meters across the UNAM CU distribution system, one in each well, in the supply points of each hydraulic sector, and one for repumping from one tank to another. There are also 210 micro water meters in the water intakes of buildings.

Remote measurement

For the remote measurement system, a protocol of communication was selected, using a radio-frequency range of 900–920 MHz. For this kind of transmission, it is necessary to instal gateway antennas that receive data from water meters every five minutes. At this point, analogue information is converted to digital information. The antennas have a coverage radius of one kilometre within line of sight. Seven antennas were installed in different parts of the campus to provide decent coverage. To enable communication between antennas and water meters, it was also necessary to instal 30 repeaters or boosters (used to receive the signal and send it to another receiver) throughout the campus.

Architecture of the telemetry system

Measurement points, data transmission, data collection, and the presentation interface are all part of the telemetry system installed on the main campus of UNAM (Figure 9). For the whole system to be reliable, it is of the utmost importance that each of these elements work properly.

Information from water meters linked to the remote lecture system is processed, analyzed and displayed using ReadCenter software (from Badger Meter). ReadCenter is a reading data management software application that provides a central location for performing various meter-reading tasks. It allows data sharing across meter-reading solutions offered by this company. The software has three components: a server (Monitor) that handles communication with the gateways; a client (Control) that defines reading schedules, performs system management, and runs reports; and a database server that stores information (Figure 10).

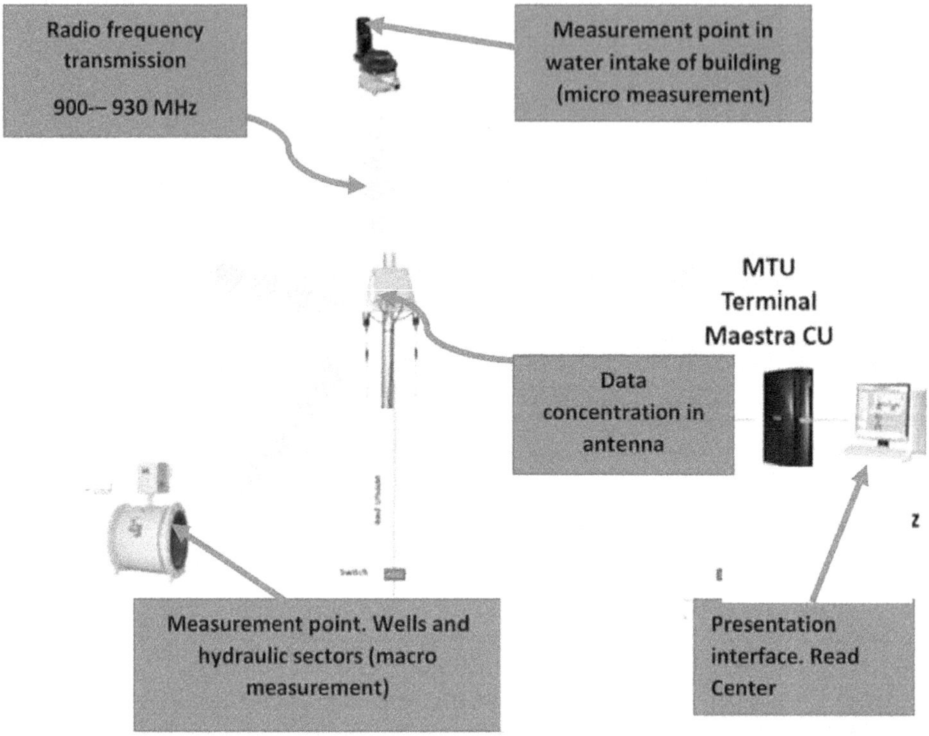

Figure 9. Telemetry system on the main campus of UNAM, including macro and micro meters, an antenna, and the server where information is displayed and analyzed.
Source: PUMAGUA (2012).

For the water meters that have not yet been linked to the system, PUMAGUA's staff make regular trips to collect the data with a portable computer called a Trimble Ranger. This is possible as the water meters have a data logger that stores information every hour. The collected data are later processed manually in Microsoft Excel.

Water quantity in UNAM's Water Observatory

To link the ReadCenter database with the PUMAGUA database, additional software was developed. It collects information from the ReadCenter database, replicates the information, and modifies its structure to make it compatible with the Water Observatory database. The latter database is administered and operated with SQL Server. The water consumption measurements collected with the portable computer are also stored in this database (Figure 10).

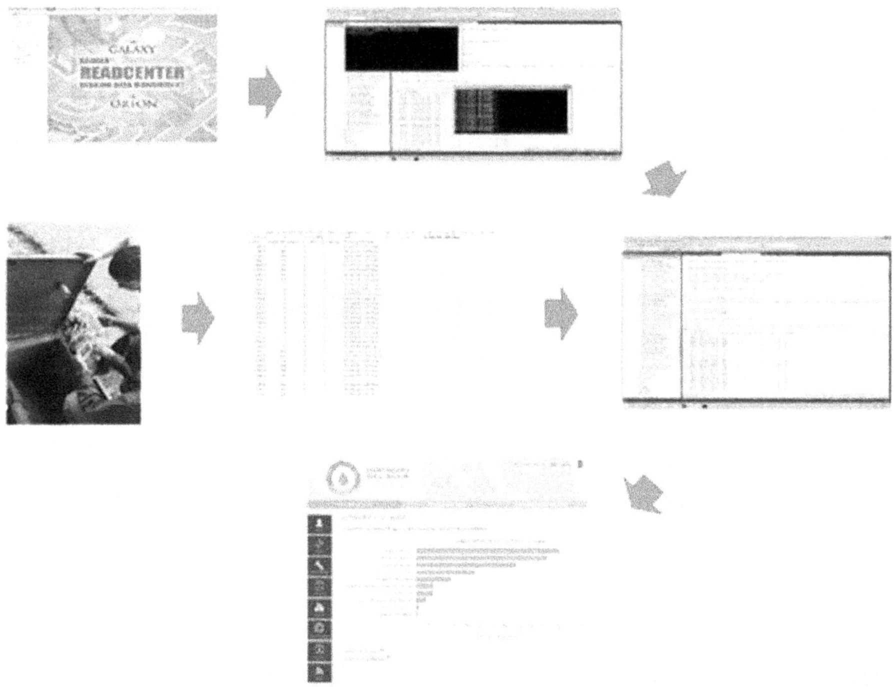

Figure 10. Integration of information into the PUMAGUA database and the Water Observatory from water meters linked to the remote system and from water meters that have not yet been linked. From left to right, top to bottom: ReadCenter programme, PUMAGUA software, collection of consumption data with portable computer, Excel file, database, water observatory.
Source: PUMAGUA (2016).

Enhancing community participation

To enhance community participation in the responsible use and management of water on UNAM campuses, several activities have taken place:

- Regular meetings with authorities of institutes, schools, and administrative offices to invite them to implement the following actions: installation of water meters, installation of water-saving toilet appliances, landscape management, installation of water fountains, attendance at PUMAGUA's workshops, and dissemination of PUMAGUA's communication campaigns;
- Activities for students and academics, such as contests for proposals to improve water management and use, water festivals, artistic contests, and water audits at schools;
- Capacity-building activities, such as workshops for maintenance staff and gardeners;
- Undergraduate and postgraduate theses conducted under the supervision of PUMAGUA;

- Popular science and indexed articles (e.g., in *Modern Environmental Science and Engineering, Revista Digital Universitaria, Revista Ciencia y Desarrollo,* and *Revista ¿Cómo ves?*); and
- Appearance in the media: newspapers, television, radio, and digital.

PUMAGUA's community participation module has two elements. The first is the implementation of the mentioned six actions suggested by PUMAGUA in the schools, institutes and administrative offices of UNAM. The platform assigns weights to each action according to their impact on water saving. For example, installing water meters and toilet appliances has a higher weight than workshop attendance. The actions are then added up, and a 'medal ranking' is applied; each entity is awarded a gold, silver or bronze medal. The purpose of this system is to give recognition to the most enthusiastic institutions and therefore encourage participation in responsible water use.

The second element is a survey designed to enable PUMAGUA to poll different university constituencies: students, professors, researchers, administrators, and maintenance staff. The survey assesses the community's perception of water management on the campus as well as their attitudes, beliefs, and behaviours regarding water use.

Publicizing the UNAM Water Observatory

To announce the Water Observatory to the Mexican public, a press conference, orga- nized by UNAM's General Directorate of Social Communication, took place in December 2016. The general structure as well as all the functions of the observatory were explained. Likewise, the observatory was included in the most important means of internal communication.

Benefits of SWM solutions within the PUMAGUA programme

Water quality

Smart technology has been highly beneficial for PUMAGUA and for the university community. It has helped the community gain confidence in drinking water quality, through enabling the immediate detection of issues in the water and providing transparency through real-time water quality data. In addition, having automated monitoring and control systems for the water on campus has increased trust within the community, as there is less potential for mistakes than with manual monitoring. The SWM technology has also enabled rapid response when a water quality parameter does not comply with Mexico's water quality regulations, so water quality can remain within drinking water requirements at all times.

Furthermore, the automation of water quality monitoring has improved campus acceptance and trust in drinking tap water. In 2012, in a face-to-face survey on the main campus, only 13% of those interviewed said they drank the tap water on campus (Espinosa-García et al., 2015). In contrast, in 2017, an online survey via the Water Observatory found that 49% of interviewees were now drinking tap water (either from the tap or from a water fountain). While care should be taken in comparing these results (as the first survey was conducted using randomly selected participants, while the second was advertised in PUMAGUA's social networks), the significant increase in willingness to drink tap water on campus indicates the benefit of continuous monitoring and display of information.

Water quantity

The use of smart technology for water quantity measurement was fundamental to gain control over the water supply and distribution system at UNAM. It permits the accurate measurement of water consumption, allowing comparisons between institutions and between different periods for the same institution. This information is then used to encourage water-saving actions across the campus. Smart technology also permits the identification of leaks and classifies them by volume, which in turn allows rapid response times for major leaks.

Water consumption measurement and leak repairs have helped UNAM institutions have consumption patterns corresponding to their characteristics, such as their population, or whether they have laboratories or gardens. Before the SWM technology and intensive leak repair, water consumption did not correspond to the usage profile of institutions.

Enablers

These project elements contributed to PUMAGUA's success at UNAM:

- Establishment of a dedicated programme and staff, with the participants deeply dedicated to PUMAGUA's goals, working even on holidays.
- The interest of specific authorities (directors, secretaries of institutes and schools). As PUMAGUA did not have funding to support the purchase of water-saving equipment, it depended on the good will of these stakeholders, who were committed to the environment and willing to invest in water-saving equipment and processes.
- Funding from the university. Without the funding from UNAM the project would not have been successful.
- National targets and policies. In particular, PUMAGUA's goal of good water quality aligned with national and local policies and laws regarding the installation of water
- fountains in schools.
- Visibility of the programme nationally and internationally. A lot of work was done to make PUMAGUA visible outside UNAM and Mexico, and this helped the project be taken more seriously by UNAM authorities.
- Multidisciplinarity. Having engineers, biologists, doctors, communication professionals, and architects, among others, working as part of the team and together on specific projects helped give PUMAGUA an integrated, flexible and adaptive vision.

Barriers

The main barrier across all elements of the project has been limited resources, including financial, human capital and technological resources.

Water quality

One of the main challenges for the project was that despite its automated system, supervision was still required, as sensors occasionally malfunctioned, requiring calibration and maintenance. This requires a significant investment, which must be budgeted for. And this has been diffcult, as the funds provided by UNAM have been reduced in recent years.

Water quantity

The Water Observatory has become one of the main tools of the water quantity area of PUMAGUA; it is therefore critical that it work properly. The PUMAGUA team checks the information displayed through the observatory on a daily basis, to detect and correct any possible errors. Some of the most frequent errors are found in consumption data. This can be caused by a water meter sending information to two repeaters, resulting in duplication of information. Another type of error has been negative measurements, which are caused by signal interruptions from repeaters, water meters or antennas. Incorrect graphs also need to be corrected immediately, as do any difficulties found when editing the information in the Water Observatory.

As a result, despite the automated technology, there is still a need to have people constantly supervising and correcting errors in the digital platform and database, as well as in the equipment installed throughout the campus. Solving this problem would require an investment in human resources which PUMAGUA cannot afford at this stage.

Another challenge is the constant updating of technology. It required significant investments to acquire the necessary technology, and recently PUMAGUA was informed that the antennas are no longer being produced, and that mobile cellular communication has become favoured over radio-frequency communication. Although a financial resources, the new technology would solve the problem of occasional data recording errors and high maintenance needs.

Lessons learned

Smart water systems cannot fully replace traditional approaches to water management. However, complementing real-time information with *in situ* water sampling and analyses enriches the end results and the validity of information. Despite the newly automated water management and monitoring system, it still requires a lot of maintenance, at least in the short term. It is important to consider this in the budget when planning SWM implementation to ensure sufficient funds for staff and the training required to maintain the SWM technology. Here is a summary of lessons learned throughout the implementation of PUMAGUA.

- Technology can change and become outdated quickly. It is important to study which technology will be the most suitable, and which new technologies are emerging that might be more effective than those currently available. For example, in the PUMAGUA case, sensors that use radio frequency were chosen, requiring additional infrastructure to ensure that the radio waves reach the entire campus. Now cellular-frequency sensors are available that do not require additional infrastructure and so may be more cost-effective in the long term.
- Care should be taken to avoid the dependence on one particular commercial product or manufacturer. It would be desirable for companies to make products compatible with each other, to provide more options. Compatibility would also likely accelerate their response time to project inquiries and issues.
- Despite the initial maintenance requirements of SWM technology, this project acts as a learning experience that will allow technology and the system to advance to a

point where maintenance issues are fixed and it becomes more robust. As with any other technology, as SWM becomes mainstream it will become cheaper and easier to implement.

- Through PUMAGUA, it was found that people are much more trusting of automated data than they are of people manually monitoring and updating data. This became evident throughout this project as people started to trust the drinking water quality when they knew the monitoring was automated, and when they could see the results for themselves on the Water Observatory. This was also observed during talks with the UNAM community, where all community members responded positively when informed that real-time automated data on the quality of UNAM water taps would be available every 5 minutes. People appreciate having access to real-time data for water quality, as it assures them of the safety of the water. Automated monitoring thus increases trust in water quality in comparison to manual monitoring, which is incapable of providing updated, accurate data every 5 minutes.
- This confidence of the public in automated monitoring could be of use to water utilities in some cities of Mexico, where water quality is adequate but people do not drink the tap water because of distrust.
- Real-time information can be very useful to build a picture of the behaviour of parameters during different periods, as it is possible to obtain continuous information. This could be useful to identify possible threats of pollution. For instance, if concentration of nitrates is gradually approaching permissible limits, this may indicate organic pollution, which could be prevented.
- PUMAGUA originally had plans to implement SWM in some of the poorer areas of Mexico City to help improve water quality and leak detection. However, it was found to be increasingly diffcult, especially in areas where basic infrastructure is limited or of very low quality, because it can be a challenge to implement effective monitoring systems where there is little to monitor. Instead, by focusing on what was feasible within the resources of the project (monitoring the houses that did have a water connection), it was possible to make a difference in these areas that provides a basis for future extension. Nonetheless, for SWM to be implemented in poor areas, at least a basic water distribution and infrastructure system must exist.

Next steps

Water quality

The next step in regard to water quality is to foster more tap water drinking through the installation of new water fountains throughout the UNAM campus. Due to project budget limitations, this will be accomplished with the help of other institutions of UNAM. The water fountains will be installed in schools, but also in social meeting places, such as green areas, sport facilities and cultural venues.

PUMAGUA will keep maintaining the real-time monitoring system and, subject to funds, this kind of system will be installed in other parts of the campus, to increase the reliability of information.

The digital platform of the Water Observatory allows connection with the systems of other UNAM campuses. Therefore, if in the near future any of these campuses acquires a real-time monitoring system, it can be linked to the platform, and information from *in situ* water quality sampling can be uploaded.

Water quantity

To upgrade the remote water consumption system at UNAM, a pilot project is planned that will allow us to decide which potential new system is the best. On the one hand, we will instal some water meters that use cellular signals instead of radio frequency. This system has several advantages; for example, it does not require antennas or signal repeaters, reducing the infrastructure costs. However, while transmitting on radio fre- quencies is free, for cellular, an annual fee must be paid to the mobile phone company, and therefore the benefits and costs of this technology need to be evaluated.

On the other hand, with the help of researchers from the UNAM Institute of Engineering, we will also try to develop our own technology for a remote water consumption monitoring system, in order to stop depending on private enterprises. This process will take longer, and will include water consumption measurement, data transmission, data reception, and uploading into the Water Observatory. Although such a strategy is time-consuming, working with a university institute is an opportunity for research and capacity building, especially for students.

Testing a sensor to detect soil moisture and to determine whether the gardens on campus are being overwatered is also being considered. If it works, several sensors will be installed throughout the green areas of the CU and linked to the Water Observatory. This will help involve the whole community, including gardeners, in the responsible use of water in irrigation.

Social participation

Information about actions recommended by PUMAGUA and implemented by schools, institutes and administrative offices will be continuously updated in the Water Observatory.

Also, the survey module of the Water Observatory will be used continuously. In particular it is of interest to monitor the ongoing progress of communication campaigns to encourage tap water drinking at UNAM, and consequently the decrease in bottled water purchases.

Conclusion

PUMAGUA was created to improve water management, use and reuse on the campuses of UNAM. Since its implementation, there have been several achievements in water savings, water quality improvement, capacity building and community participation. SWM technology has been fundamental to these achievements. It has helped produceand disseminate a substantial amount of information, and to convince the community of the validity of this information, which it sees as free from human error.

Updating the SWM technology would be an opportunity to extend it to other aspects of monitoring and maintenance, for instance, the installation of soil moisture sensors. It is also an opportunity to develop our own technology, to reduce dependence on commercial providers.

This technology does not work independently from human beings, however: it needs human resources for updating and for the correction of errors. Also, SWM must be supplemented by other sources of information, such as on-site sampling. In the case of PUMAGUA, real-time sensors are located in only one place and provide information in six water quality parameters. In contrast, Box 2 our on-site water sampling allows us to monitor 75 drinking water fountains, 25 intakes and 52 water storage tanks. Moreover, water sampling allows us to analyze the 47 parameters covered by Mexican regulations, which cannot be monitored by real-time sensors.

Disclosure statement

No potential conflict of interest was reported by the authors.

References

ALDF. (2013). *Buscan instalar bebederos en escuelas públicas*. Legislative Assembly of the Federal District. http://www.aldf.gob.mx/comsoc-buscan-instalar-bebederos-escuelas-publicas–12864.html

Capella. (2015). *En México se pierde 40 por ciento del agua potable por fugas en redes: experto de UNAM*. (Press reléase). http://www.iingen.unam.mx/es-mx/difusion/Lists/t.b./DispForm.aspx?ID=377

Capella-Vizcaino, A., Vega-Serratos, E., & Herrera-Alanis, J. L. (2008). *Programa de largo plazo para el abastecimiento de agua potable para la Zona Metropolitana del Valle de México*. Informe Final para la Comisión Nacional del Agua.

CONEVAL. (2017). *Medición de pobreza*. http://www.coneval.org.mx/Medicion/Paginas/PobrezaInicio.aspx

DOF. (2016). *ACUERDO número 27/12/16 por el que se emiten los Lineamientos de Operación del Programa de la Reforma Educativa*. http://www.dof.gob.mx/nota_detalle.php?codigo=5468071&fecha=29/12/2016

Espinosa-García, A. C., Díaz-Ávalos, C., González-Villarreal, F. J., Val-Segura, R., Malvaez-Orozco, V., & Mazari-Hiriart, M. (2015). Drinking water quality in a Mexico City University Community: Perception and preferences. *Ecohealth*, *12*(1), 88-97. doi: 10.1007/s10393-014-0978-z.

Freshwater Action Network. (2017). *México ahora más grande consumidor de agua embotellada*. http://www.freshwateraction.net/es/content/m%C3%A9xico-ahora-m%C3%A1s-grande-consumidor-mundial-de-agua-embotellada–1

Geo-Mexico. (2017). *Mexico's seven climate regions*. http://geo-mexico.com/?p=9512

González Villarreal, F. J., Rodríguez Briceño, E., Padilla Ascencio, E., & Lartigue Baca, C. (2015). Percepción del servicio y cultura del agua en México. *H2O: Gestión del agua, 7*.

HACH Company. (2006). *User manual* (4th ed.).

Instituto Nacional de Estadísitica y Geografía. (2010). *Censo Nacional de Población*.

Instituto Nacional de Estadísitica, Geografía e Informática. (2002). *Cuaderno Estadístico de la Zona Metropolitana de la Ciudad de México* (Edición 2002).

National Autonomous University of Mexico. (2017). *Acerca de la UNAM*. https://www.unam.mx/acerca-de-la-unam/unam-en-el-tiempo/cronologia-historica-de-la-unam/1950

Organization for Economic Cooperation and Development. (2015). *Estudios territoriales de la OCDE*. https://www.oecd.org/regional/regional-policy/valle-de-mexico-highlights-spanish.pdf

Organization for Economic Cooperation and Development. (2017). *Inequality*. http://www.oecd.org/social/inequality.htm

Orta de Velásquez, M. T., González Villarreal, F. J., Yañez-Noguez, I., Val Segura, R., Lartigue Baca, C., Monje-Ramírez, I., & Rocha Guzmán, J. D. (2013, October 22–25) Implementation of efficient use and water quality control within PUMAGUA programme. In *7th IWA international conference on efficient use and management of water (efficient 2013) Paris, France*.

PUMAGUA. (2008). *Diagnosis*. National Autonomous University of Mexico. http://www.puma gua.unam.mx/assets/pdfs/informes/2009/diagnostico_2008.pdf
PUMAGUA. (2012). *Annual inform* [unpublished].
PUMAGUA. (2016). *Annual inform* [unpublished].
Quadratin. (2017). *Urgen políticas para reducir consumo de agua embotellada*. https://www.quad ratin.com.mx/sucesos/urgen-politicas-reducir-consumo-agua-embotellada/
Romero Sánchez, G. (2015, July 25). Un millón de capitalinos carece de agua de calidad; 6% del abasto no es potable. *La Jornada*. http://www.jornada.unam.mx/2015/07/25/capital/028n1cap
Romero-Lankao, P. (2010). Water in Mexico City: What will climate change bring to its history of water-related hazards and vulnerabilities? *Environment & Urbanization*, *22*(1), 157–178. https://doi.org/10.1177/0956247809362636
Secretaría de Salud. (2016). *Cuarto Informe de Labores, 2015–2016*. https://www.gob.mx/cms/uploads/attachment/file/131363/4to_Informe_de_Labores_SS.pdf
Torregrosa, L. (2012). Los recursos hídricos en México: Situación y perspectivas. In J. P. Laclette & P. Zúñiga (Eds.), *Diagnóstico del Agua en las Américas* (pp. 309–354). Foro Consultivo Científico y Tecnológico.
World Bank. (2017). *GDP growth*. https://data.worldbank.org/indicator/NY.GDP.MKTP.KD.ZG

K-water's Integrated Water Resources Management system (K-HIT, K-water Hydro Intelligent Toolkit)

Sukuk Yi, Munhyun Ryu, Jinsuhk Suh, Shangmoon Kim, Seokkyu Seo, Seonghan Kim and Sungphil Jang

ABSTRACT

This study details the Hydro Intelligent Toolkit (K-HIT) developed in 2002 by K-water, the Korean water agency, to cope with extreme events such as floods and droughts. It is structured to outline the challenges to water management in Korea, the development and components of K-HIT, its effectiveness in practice, and lessons learned.

Summary

Water management is a challenge in Korea due to limiting geographical features, such as short watercourses, and high rainfall variability across the regions and seasons. Korea also faces regular water-related disasters such as extreme flooding and droughts, which are increasing in intensity due to the change in climate, thus creating increased necessity for national water management and security.

To address this, Korea has made great efforts to resolve temporal and regional variability through the construction of multipurpose dams and multiregional water supply systems. This investment in water resources in Korea has increased the water supply for industrial and domestic use, alongside supporting Korea's national economic development.

K-water is responsible for managing floodwater and for supplying water through the operation of water resources infrastructure, including 34 multipurpose and water supply dams, four flood control dams and reservoirs, 16 weirs, and one estuary barrage (similar to a low dam wall). Of these, the multipurpose dams operated by K-water account for 62% of total dam supply and 94% of flood control capacity.

In order to protect people from drought and flood disasters through more efficient water resource management, K-water has constructed a scientific river operation system which links the rivers in connecting watersheds. The aim of this system is to implement Integrated Water Resources Management technology in rivers for the purpose of increasing water quantity and improving water quality concurrently.

In 2002, the K-water Hydro Intelligent Toolkit (K-HIT) was introduced to provide an integrated water management system based on information and communication technology (ICT). K-HIT has five functions: real-time hydrological data acquisition, precipitation forecasting, flood analysis, reservoir water supply, and hydropower generation. By using K-HIT, K-water can minimize flood damage by storing more water during the flood season and prevent droughts by supplying stored water during the dry season.

Through the introduction of K-HIT, K-water has been able to effectively deal with floods that occurred in 2012, 2013 and 2015. In this way K-HIT also contributes to the achievement of the UN's Sustainable Development Goals, including Goal 6 (availability and sustainable management of water) and Goal 11 (making cities inclusive, safe, resilient and sustainable), by preventing disasters such as floods and securing water in droughts.

Background

Climate and water management characteristics of Korea

Topography and precipitation features

Water management in Korea is broadly divided into six zones: the Han River in the Seoul metropolitan area and Gyeonggi, the Geum River in Chungcheong, the Seomjin River and the Yeongsan River in Jeolla, the Nakdong River in Gyeongsang, and the Jeju and Ulleung Islands (Figure 1).

As shown in Figure 2, Korea experiences highly concentrated rainfall in the summer (June to August), while precipitation is very scarce in the winter (October to January). Rivers in Korea experience severe fluctuations in flow, with the rainy season in particular creating an extreme challenge for water management due to the high flood levels and runoff. The dry season results in very low river levels, which creates its own water management challenges. Table 1 shows a comparison of flow variation coefficients between major rivers in Korea and in other countries.

An analysis of the patterns of precipitation change, as shown in Figure 3, indicates that the average rainfall over a 10-year period is increasing (by approximately 4% since the 1970s), and rainfall variability has been gradually increasing from the 1990s. The annual average rainfall over the past 100 years shows great variability, from a minimum of 754 mm in 1939 to a maximum of 1756 mm in 2003. In consideration of these natural precipitation characteristics, Korea has made great efforts to resolve temporal and regional variability through the construction of multipurpose dams and multiregional water supply systems.

Furthermore, weather anomalies such as extreme floods and droughts, which seem to be intensifying due to climate change, are occurring more frequently. Recently, Korea has been suffering from a severe drought. Annual rainfall in 2013 was 11% lower than the average annual rainfall, and in 2014 it was 10% lower (Table 2). By August 2015, the annual rainfall was only 64% of the historical average. Due to the increasing frequency of extreme weather conditions, Korea faces many challenges in terms of managing water resources and ensuring water security.

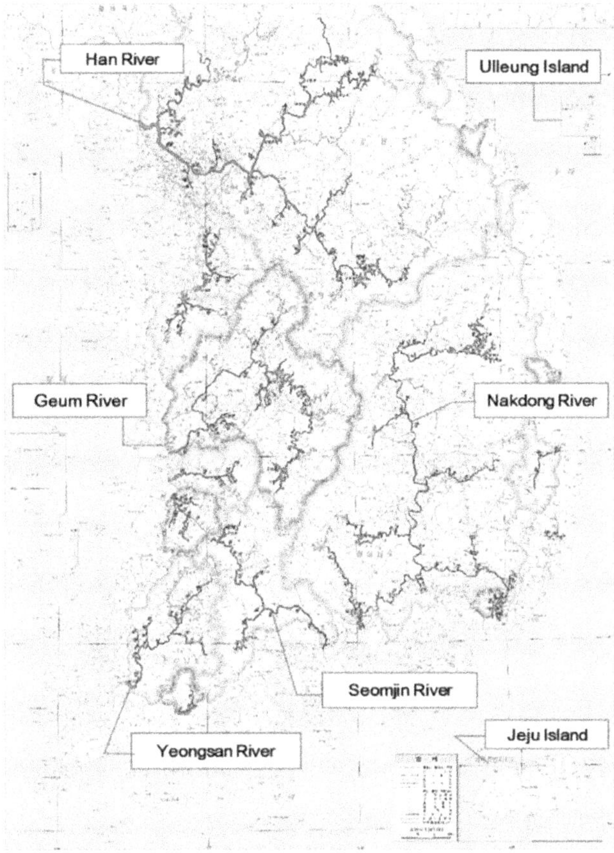

Figure 1. The six water management zones of Korea.

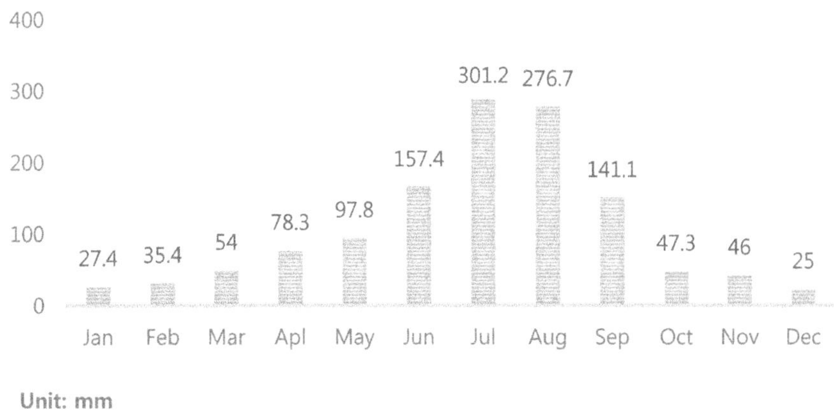

Figure 2. Monthly rainfall in Korea.

Water resources availability

The average annual rainfall in Korea is 1274 mm (calculated between 1973–2011), 1.6 times the world average precipitation. However, due to the high population density in

Table 1. Flow variation coefficients in rivers.

River	Flow variation coefficient	River (country)	Flow variation coefficient
Han	90 (390)	Tone (Japan)	115
Nakdong	260 (372)	Seine (France)	34
Geum	190 (300)	Rhine (Germany)	18
Seomjin	270 (390)	Nile (Egypt)	30
Yeongsan	130 (320)	Mississippi (USA)	3

Note: Figures in parentheses refer to coefficients before multipurpose dams were constructed.
Source: Ministry of Land, Infrastructure and Transportation of Korea (MoLIT) (2017), Fourth long-term comprehensive plan for water resources (2001–2020), 3rd revision.

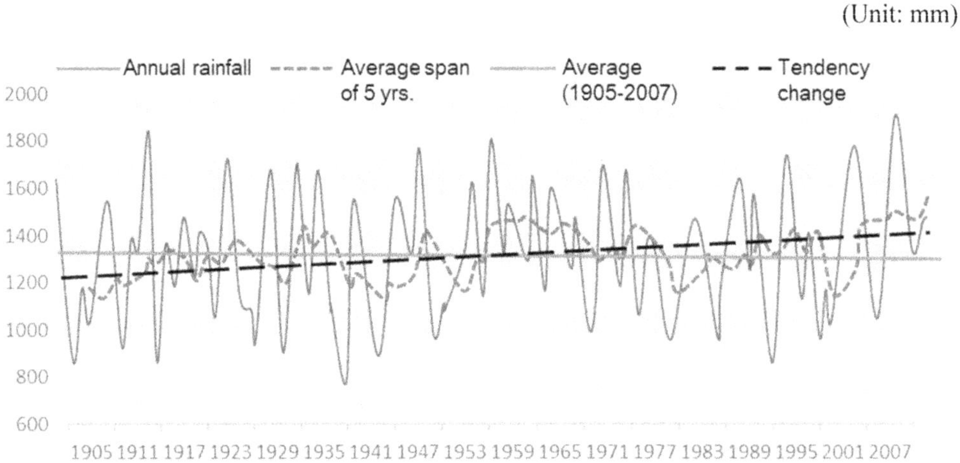

Figure 3. Average rainfall in Korea, 1905–2007 (mm).
Source: MoLIT (2017), Fourth long-term comprehensive plan for water resources (2001–2020), 3rd revision.

Table 2. Annual rainfall in Korea, 2006–2015.

	2006	2010	2011	2012	2013	2014	2015.8
Rainfall (mm)	1,424.3	1,444.9	1,622.6	1,479.1	1,162.9	1,173.7	659.4
Percentage of historical average	109%	110%	124%	113%	89%	90%	64%

Source: MoLIT (2017), Fourth long-term comprehensive plan for water resources (2001–2020), 3rd revision.

Korea, the average annual rainfall per capita is only 2660 m^3, approximately 16% of the world average. Per year, Korea has a total water supply of 132.3 billion m^3. In 2007, the total amount of water use was 37.2 billion m^3 (dams, 20.9 billion; groundwater, 4.1 billion; rivers, 12.2 billion), which accounts for 28% of the total amount of water resources available. Total water use is 1.8 times the normal water runoff (runoff that occurs outside the rainy season, 21.2 billion m^3), and thus flood runoff (runoff that occurs during the rainy season of June to September) is reserved in impoundments such as dams and other reservoirs to be used when necessary. The total domestic, industrial and agricultural water use is about 25.5 billion m^3 per year, which is approximately 34% of the total available water resources.

Water use by purpose

Agricultural water accounts for the largest proportion (48%) of the total 25.5 billion m^3 per year of water used, followed by domestic (23%), river maintenance (23%), and industrial use (6%). Notably, river maintenance flow continues to increase over time.

The abstraction rate for domestic, industrial and agricultural purposes stands at 34%, confirming that Korea is a 'water-stressed' country (Table 3). Hence there is greater water insecurity, and it is more challenging to manage water quality and conserve river ecosystems.

Flood risks

It is essential to take pre-emptive measures and preparation in order to mitigate the effects of natural disasters, given that weather anomalies and vulnerability driven by climate change are aggravated in Korea.

Approximately 70–80% of the total annual rainfall is concentrated in the summer, making Korea highly vulnerable to water-related disasters. The average annual damage caused by water-related disasters over the last 10 years is estimated at KRW2 billion (USD1.9 billion). Korea's flood risk index is much higher than other countries, resulting in the need for systematic and proactive responses in order to prevent extensive damage from disasters and climate change.

Population and water supply rate

The population of Korea increased from 31,435,000 in 1970 to 52,419,000 in 2014. In that time water has been supplied to an increased proportion of the population, from 33.2% (10,430,000 people) in 1970 to 96.1% (50,373,000 people) in 2014 (Table 4).

National water resources management system

Korea has 17,759 dams in total. These dams supply 18.8 billion m^3 of water annually, 56% of Korea's annual water use of 33.3 billion m^3, and they control floods with a capacity of 5.3 billion m^3. Of these, the multipurpose dams operated by K-water account for 62% (11.7 billion m^3) of total dam supply in Korea and 94% (5 billion m^3) of flood control capacity.

Table 3. Water stress level by country.

Abstraction rate	Water stress	Country
Below 10%	Low	New Zealand, Canada, Russia
10–20%	Middle	Japan, USA, UK, France, Turkey
20–40%	Mid-high	Korea, China, India, Italy, South Africa
40%	High	Iraq, Egypt

Source: MoLIT (2017), Fourth long-term comprehensive plan for water resources (2001–2020), 3rd revision.

Table 4. Population and water supply.

Year	1970	1975	1980	1990	2001	2011	2014
Population (thousands)	31,435	34,709	38,124	43,520	48,289	51,717	52,419
People receiving water supply (thousands)	10,430	14,961	20,809	33,631	42,402	48,938	50,373
Percentage of population with water supply (%)	31.2	43.1	54.6	77.3	87.8	94.6	96.1

Source: Choi Han-ju, 'Analysis of the Effects of Water Resources Development in Korea's Economic Development Process', December 2016.

The flood control process of multipurpose dams is based on the rainfall data from the hydrological data management system, connected with the weather information system and the real-time online network of the river basins managed by K-water. K-water also performs rainfall-runoff analysis through the flood analysis system. This system is managed to make full use of the flood control capacity by determining the appropriate controlled discharge amount and timing. To do this K-water considers the capacity of the downstream dams and conduct flood control of the entire water system in consultation with related organizations, such as the Flood Control Office.

K-water operates and manages important water resources infrastructure, including 34 multipurpose and water supply dams, four flood control dams and reservoirs, 16 weirs and one estuary barrage (Figure 4), all of which are government assets.

The basins managed by K-water account for 48% of the nation's total land area, 95% of the nation's entire flood control capacity and 65% of the total water use. As the number of

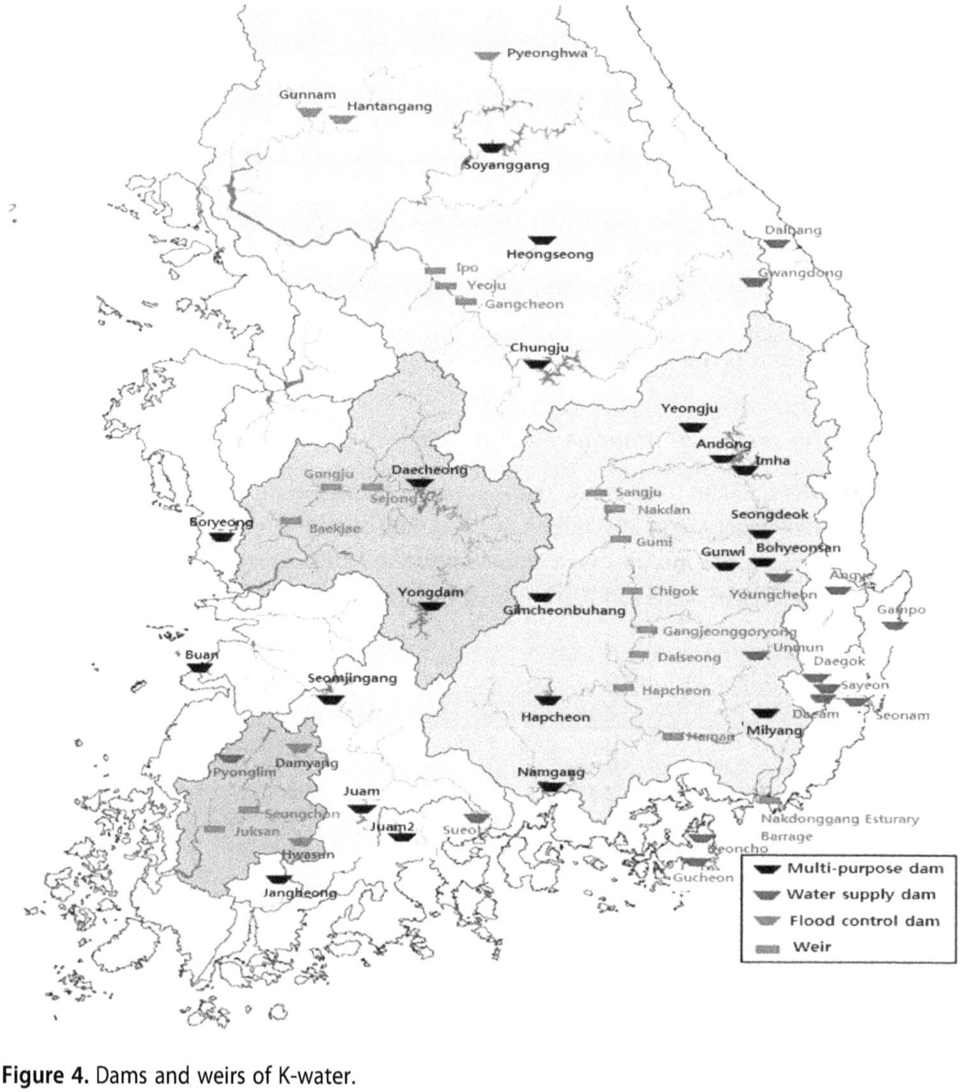

Figure 4. Dams and weirs of K-water.

dams K-water manages continues to increase, K-water needs to consider ways to combine dam operations to effectively conduct water resources management in river basins.

The challenge

Increasing frequency and severity of water-related disasters

The frequency of severe floods and droughts in Korea has increased in recent years due to climate change. Although the size of this phenomenon is difficult to measure due to a broad range of uncertainties, the increase in water-related disasters such as intensive precipitation events and irregular drought cycles is a significant and urgent issue that Korea is now facing.

The financial cost of the damage caused by natural disasters over the past 10 years in Korea, 87% of which were water related, was KRW1.7 trillion (USD1.55 billion) per year, 5.3 times the cost in the 1980s and 3.1 times the cost in the 1990s. The annual restoration cost was even higher (approximately KRW2 trillion, USD1.8 billion). In addition, as 62% of dam facilities are over 30 years old, water management hazards are rapidly increasing, creating a further serious water management and safety issue.

For example, in 2012, for the first time since 1962, four typhoons struck the Korean Peninsula (including three in close succession at the end of the rainy season), causing precipitation almost twice the monthly average from the middle of August to the middle of September. In 2013 the rainy season was the longest on record (weather observation data were first collected in 1904), with heavy rain in the middle region of Korea, while insufficient rain in the southern region resulted in a regional deviation of over 200 mm. The worst impacted was Jeju Island, which experienced approximately USD300 million in drought damage, for the first time in 90 years. According to the Fourth Assessment Report of the Intergovernmental Panel on Climate Change, it is expected that the risks of flooding and drought disasters in high-latitude countries in the northern hemisphere (including Korea), as well as the intensity and variation in rainfall, will continue to increase significantly.

Significant change of water management conditions after the Four Major Rivers (Restoration) Project

In addition, water management conditions in Korea changed significantly as a result of the Four Major Rivers (Restoration) Project, which was completed in 2012. K-water used to control floods and water level with linked operation among more than 30 multi-purpose and water supply dam and river stations. The project's changes to river facilities (e.g., weirs and river-crossing sections) affected the streamflow of the river, increasing the consideration factors for flood control and algae bloom.

Smart Water Management solution

Input

The (smart) Water Management Centre for the integrated operation of dams and weirs

In 2002, K-water established the Water Management Centre, with 39 experts on water resources, including water management, ICT, power generation and weather. With the opening of a comprehensive water control room, K-water greatly enhanced its water resource management capabilities by integrating the operation of dams. The Water Management Centre expanded its management staff to 55 people in 2010 and 67 in 2016 in an expansion of management facilities and functions.

The Water Management Centre performs the following stages of work during normal and flood times.

- Normal stage: one person is on duty after working hours (18:00–09:00) if there are no forecasted weather issues.
- Caution stage: If the national rainfall forecast requires the operation of dams and weirs for flood control, eight people will operate the facilities (in groups of four).
- Alert stage: If weather report is issued and water discharge is needed to carry out flood control, 13 people will operate (in three groups).
- Serious stage: If nationwide floods and massive flood damage occur, 24 people (split into two groups) will be operate the system.

The Flood Control Capacity Enhancement Project

The Flood Control Capacity Enhancement Project on existing dams is meant to prevent disasters from excessive flood due to changes in the climate and thus protect lives and property by increasing the safety of dams.

To prepare for the increasing frequency of heavy rain due to recent climate changes, the standard for the design flood has been changed from frequency of flooding (100 y or 200 y) to probable maximum flood.[1] This is the amount of flooding that results from the maximum possible amount of water, which is the theoretical maximum precipitation that can physically occur in a given period of the year for a given watershed area at a given location over a given duration. Therefore, the design flood has been enlarged due to this change.

The amount of inflow into the dams is much higher than the design capacities when the dams were initially constructed. Thus, flood control capacity enhancement is required to guarantee the permanent safety of dams. The types of measures include spillway expansion, auxiliary spillways and parapet walls.

The Flood Control Capability Enhancement Project for 24 dams, begun in 2003, is still in progress. Seventeen dams have been completed (as of 2017), three are under construction, and four are being planned.

Establishment of operational regulations for dam-weir operation

In order to integrate the water facilities' operation with the existing dam management regulations, regulations on linked operation between dams and weirs have been developed to ensure that the water resources facilities can be more efficiently managed.

Scientific and systematic operation is implemented to both secure additional water resources for the future and to allow water facilities to pre-emptively discharge water for the purpose of flood control. For the management of weirs, it is necessary to consider river inflow situations, discharge situations of upstream dams, and river water quality.

Drought Response Guidelines

Step-by-step 'Drought Response Guidelines' have also been developed for dam operators to enable them to pre-emptively prepare for drought situations and to carry out the measures required to manage the water supply. The guidelines ensure that the proper water supply is provided during droughts in order to minimize any social impacts, including public inconvenience.

The 'standard' daily storage volume required for the dams to stably supply water for the next year is set in the guidelines. If at any time the actual storage volume is lower than the standard volume, the amount of water discharged from the dams is reduced in stages in accordance with the advice of relevant agencies.

- Ordinary season: Water supply reductions are flexibly implemented to supply water through conjunctive dam operation in the river basin based on the capacity of each dam.
- Shortage of water supply: If conjunctive dam operation is unable to satisfy the water demand, dam water supply is reduced in order to prevent the interruption of daily or industrial water supply in the future.

In Korea, monthly operation plans for multipurpose dams are confirmed by the prior approval of a joint Collaborative Operation Committee with Dams and Weirs.

The drought response stages and major action plan in each stage are:

(1) Notice stage: adjustment of water supply to contracted quantity, real-time monitoring and progress sharing with relevant agencies.
(2) Caution stage: reduction of river maintenance water, support water supply to the public, such as bottled water, and campaign of water saving by media.
(3) Alert stage: Reduction of irrigation water and structural countermeasures, such as development of underground wells and leakage reduction.
(4) Serious stage: Restrict domestic and industrial water supply and develop sustainable countermeasures, including building new dams and diversifying water sources.

The standard storage volume is the minimum volume required to supply water in accordance with the drought stage, as estimated by trial and error methods and considering safety of the water supply (95%, 475 out of 500 years; Figures 5 and 6).

If a dam storage approaches the standard storage volume, the reduction of water supply will be implemented in accordance with the drought response guideline after approval of the Collaborative Operation Committee with Dams and Weirs. The order of reduction is primarily river maintenance, then irrigation, then domestic and industrial water (Table 5).

If the water supply reaches the caution stage during the reduction process, the reduction will continue until it exceeds the normal stage (with exceptions at the request of the relevant agencies). If the dam storage exceeds the normal stage, the water supply should be recovered up to the contracted water quantity volume in consultation with the relevant organizations.

The development of the IWRM system

Korea relies on the supply of surface water by rivers and dams for 90% of the total annual water resources supply (approximately 33.7 billion m^3). During most of the year (except

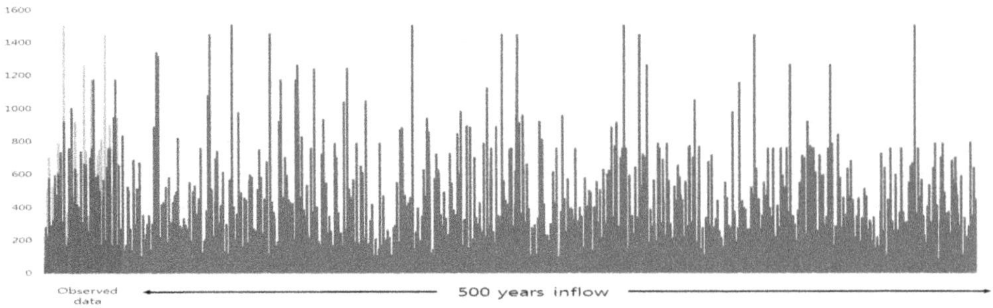

Figure 5. Estimation of long-term inflow for 500 years by stochastic analysis using SAMS model. This is an extension of inflow maintaining statistical characteristics using the observed data.
Source: K-water.

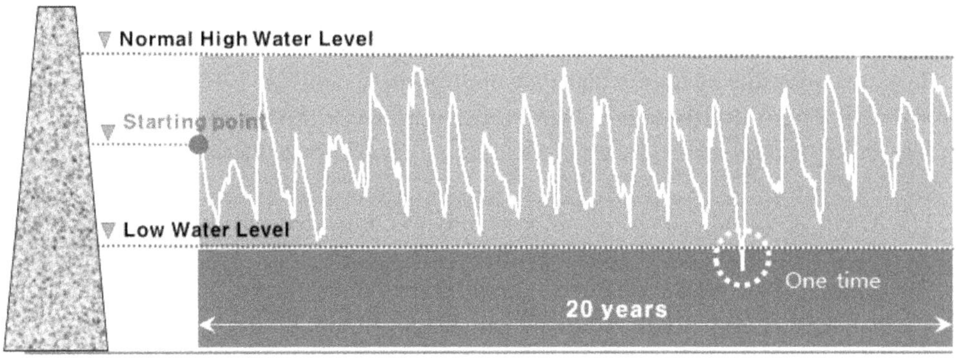

Figure 6. 95% safety for water supply (1 year shortage out of 20 years).
Source: K-water.

Table 5. Required reduction in stages.

Stage	Required reductions
Notice	Surplus water (only contracted water is provided)
Caution	River maintenance water
Alert	River maintenance water + irrigation water
Serious	River maintenance water + irrigation water + (domestic and industrial water) ratio

the flood season), a number of major rivers, including the Han, are controlled by the supply volume of multipurpose reservoir groups in the upper-middle area of the basin. Therefore, in order to protect the local people from drought and flood disasters through more efficient water resource management, it is essential to construct a multipurpose dam with a reservoir and a scientific river operation system in a watershed or a metropolitan watershed which links the rivers.

K-water has constructed this system in three steps with the aim of implementing Integrated Water Resources Management in rivers for the purpose of increasing water quantity and water quality at the same time.

- Phase 1 (2001–2004): Development of the Real-Time Watershed Management System, a database-based modularized analysis system consisting of simulation models of reservoir groups and watershed runoff, and an Integrated Water Resources Management System.
- Phase 2 (2004–2007): A geographic information system and a real-time Reservoir Turbidity Monitoring and Modelling System implemented in parallel with the calibration of the elemental technologies developed in Phase 1 by applying the Geum River and Nakdong River basins.
- Phase 3 (2007–2011): Research and business development technology developed through generalization and commercialization of various technologies and systems for analyzing water quality and to improve the water resource management capacity of dams and rivers connected in a watershed, with consideration of water ecology and ecosystems.

The K-water Hydro Intelligent Toolkit

The main aim of this project was to maximize efficient water supply and to minimize flood disasters. To achieve this K-HIT, an integrated water management system based on ICT, was developed. The multiple functions of this system can be operated by anyone, anywhere, at any time. This system is also sharable and smartphone compatible. K-HIT has five sub-systems: real-time hydrological data acquisition, precipitation forecasting, flood analysis, reservoir water supply, and hydropower generation (Figure 7).

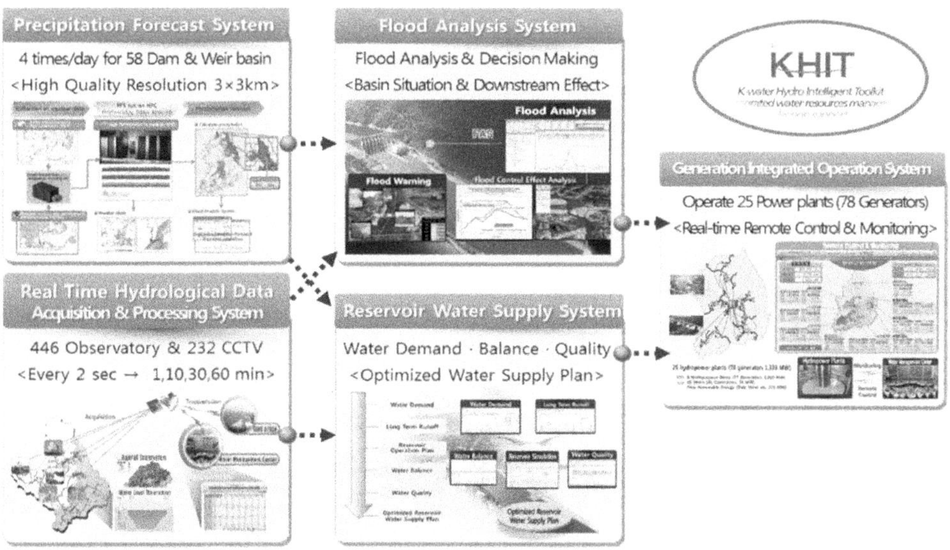

Figure 7. K-water's water resources management procedures based on K-HIT.
Source: K-water.

The K-HIT procedures are applied in several steps to operate the dams, weirs and hydropower plants. First the data is gathered and processed from our observatories using the Real-Time Hydrological Data Acquisition and Processing System (RHDAPS). Then we forecast the rainfall for dam and weir basins using the Precipitation Forecasting System (PFS). These observed and forecasted data are used for flood control and water supply. Then, based on the results of the Flood Analysis System (FAS) and the Reservoir Water Supply System (RWSS), the volume of dam discharge, the gate opening time, and withdrawal of the water from dams are determined. Finally, the Generation Integrated Operation System (GIOS) is used to efficiently operate the hydropower generation facilities to consider the discharge from the dams and weirs in the water management operation.

Precipitation Forecasting System

The PFS (Figure 8) was developed by K-water to forecast the average precipitation of the dam and weir basins. The Korea Meteorological Administration and the US National Oceanic and Atmospheric Administration provide the initial data, including temperature, humidity and wind speed, which are then used as input data for K-PPM, a 3 km × 3 km high-spatial-resolution forecasting model.

Using a high-performance computer, the PFS provides forecasts for the next five days (updated four times a day) and makes various weather maps, including rainfall prediction. In addition, it analyzes the predicted rainfall for each basin, with the resulting data applied to the flood analysis model.

Real-Time Hydrological Data Acquisition and Processing System

All data related to dam operations and precipitation, water level and hydraulics are saved in the database server at the Water Management Centre and are managed by the RHDAPS

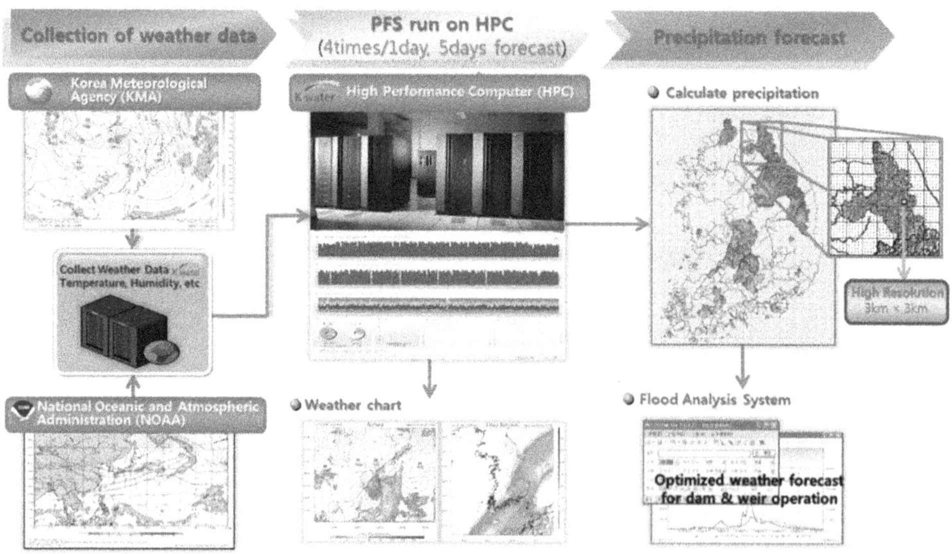

Figure 8. Precipitation Forecasting System or dam and weir operation.
Source: K-water.

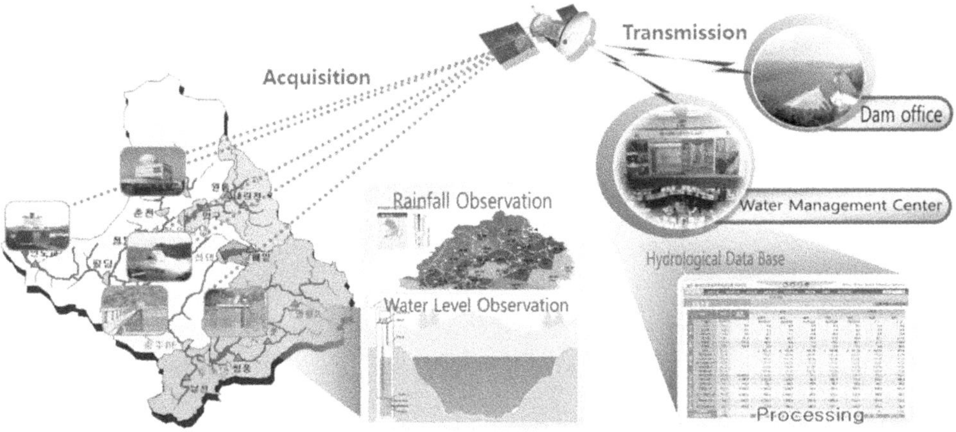

Figure 9. Real-Time Hydrological Data Acquisition and Processing System.
Source: K-water.

(Figure 9). Real-time hydrological data such as rainfall, water level, inflow and outflow are collected every two seconds from over 400 observatories. The RHDAPS then produces user-friendly operational data at 10-minute and 30-minute intervals. As such, this system familiarizes the operators with key data on inflow, outflow and the operational status of the hydro-structures. To increase the safety and reliability of data gathering, a dual communication network has been established using both satellite and CDMA.[2]

Flood Analysis System

K-water built the FAS for flood control and dam and weir operation. The FAS performs flood analyses using rainfall forecasting data provided by the PFS and hydrological information obtained through the RHDAPS. As the first step, the system decides the operation methods for each reservoir through rainfall-runoff analysis using storage function and hydrological channel tracking. Next, the PFS operates a joint operation model for the reservoir system to prevent flood damage by minimizing peak flow downstream. At this time, discharge is decided upon using optimization techniques. Simulations are performed to determine the optimum discharge plan, with consideration of the field site constraints. After this simulation, the discharge plan is revised, and the hydraulic model is operated by applying the reservoir operation plan and tide level. After assessment of whether the flooding of rivers will occur, a final discharge decision is made. Through these processes, an optimal release plan is realized (Figure 10).

After determining the optimal release schedule, the dam's discharge schedule is disseminated to the local government, broadcasting companies, and people who live in the downstream area of the dam, at least three hours before the gates are opened. After completing the flood control measures, we also evaluate and communicate the result of the release plan to make people aware of the positive effects of the dam operation.

The rainfall-runoff simulation in the FAS is applied to the storage function model, which is a nonlinear and concentrated model. A one-dimensional numerical model

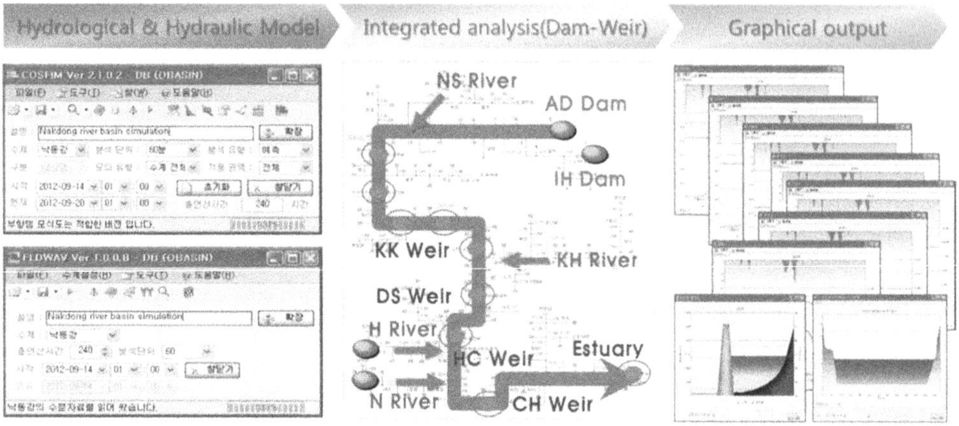

Figure 10. Flood Analysis System.
Source: K-water.

considering dynamic waves is also applied to the hydraulic analyses in the FAS. Using these models we are able to operate a detailed simulation of coordinated dam–weir operations.

Reservoir Water Supply System

Water management in the dry season is based on a system for stable water supply during an unexpected drought.

The RWSS (Figure 11) was developed to make optimal water supply plans from reservoirs considering water demand, water balance and water quality. First, long-term rainfall-runoff is analyzed for the entire basin, including the consideration of long-term predicted rainfall. Then, medium- and long-term operation plans are made based on analyzed runoff situations and water demand. With this operation

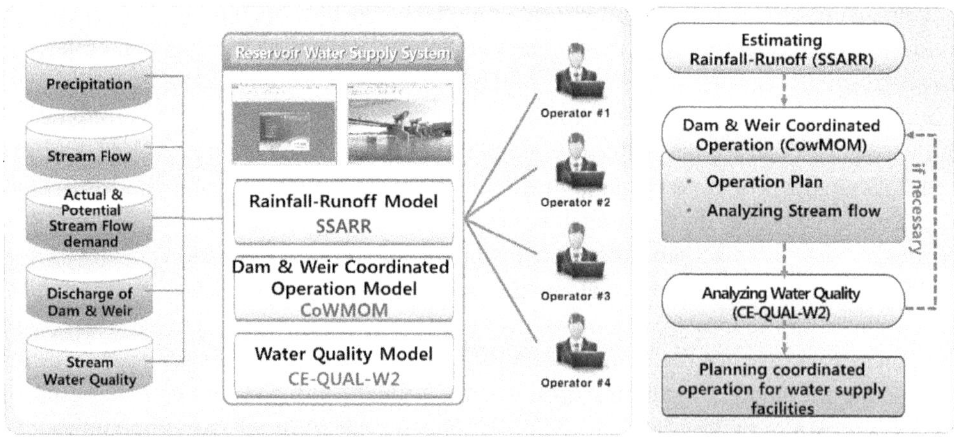

Figure 11. Reservoir Water Supply System.
Source: K-water.

plan, water quality in rivers and weirs is examined, and if it does not meet the criteria, the operation plan is re-produced. Once the water quality meets the criteria, the plan is given to the joint operation council after consultation with inter-agency organizations. The council then comes to a decision based on the consideration of general water resources use in entire basins and operation situations, and the final operation plan for water resources facilities for stable water supply is completed. For reservoirs, the operating plan by the RWSS is checked by water level standards. Guidelines are set in advance for unexpected severe droughts, with regular severe droughts anticipated every 20 years.

By applying the sequent peak method (a method used to determine the reservoir capacity required to meet demand over a given period), monthly water level guidelines are calculated to supply water stably, taking into consideration the design drought inflow (expected inflow in drought). If water levels in the guidelines are higher than the restricted water levels of reservoirs, inflow and water supply resume.

Generation Integrated Operation System

K-water developed the GIOS (Figure 12) in 2005 to enhance competitiveness in the electric power market. Using the GIOS, 25 hydropower plants (and 78 generators) are remotely operated and monitored from the Water Management Centre on a real-time basis. While hydroelectric energy is a small part of the total amount, it contributes to the stable electricity supply, since it plays a pivotal role in managing peak demand situations. In peak demand, the GIOS automatically reacts to ensure energy demands using hydropower. To ensure that the system is protected against breakdown or cyberterrorism, the GIOS has a failover function and security system, alongside dual servers and networks to ensure continuous service, and an antivirus program and surveillance cameras for system security.

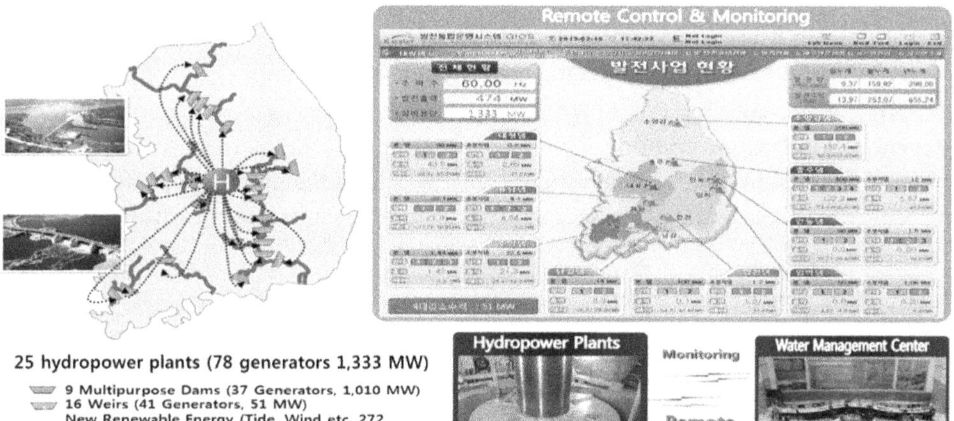

Figure 12. Generation Integrated Operation System.
Source: K-water.

Outcomes and impact

Operation of dams against floods

Flood control in 2012 and 2013

In June 2012, the rainy season started later than usual (July instead of June) and finished earlier than usual. After it ended, precipitation was significantly higher than usual due to the passing of continuous troughs of low pressure and three consecutive typhoons, which occurred for the first time in meteorological observation history. Accumulated precipitation nationwide in multipurpose dam basins in 2012 was 1402 mm, approximately 110% of the historical average, 967 mm of which occurred during the rainy season (Table 6).

Due to the close cooperation system between the upstream and downstream facilities, K-water only needed to release 29% of the total inflow through the floodgates during this period, which played a major role in the reduction of flood damages. In other words, K-water stored most of the inflow in dams and only released water when the downstream flood discharge decreased, so the water level at the main point downstream decreased to the required maximum of 5.1 m.

Overall, K-water minimized the flood damages through scientific and effective dam operation, storing 0.6 million m^3 of water during the rainy season and discharging 0.8 million m^3 of the total inflow (1.4 million m^3). Discharge in this period (excluding the amount required for water supply, instream maintenance and irrigation) was approximately 38% of the total discharge and 23% of the total inflow. This approach contributed to ensuring stable flood control as well as securing the storage water needed for supplying water during the dry season and for improving the quality of the water quality supplied.

In 2013, despite only 80–90% of the annual average precipitation occurring in most basins (with the exception of the Han River basin), the Geum River basin secured the average annual level of water storage, and other basins secured more than the average, due to K-water's previous efforts to secure additional water in preparation for insufficient rain. As a result, K-water secured 119% of the average water storage and 105% of the goal at the end of rainy season. By storing water prior to the rainy season K-water also decreased the water level by 3.2 m in the Han River, 2.1 m in the Nakdong River, 1.2 m in the Geum River and Seomjin River (Table 7), reducing the water discharge during the rainy season and thereby reducing the potential for flooding.

Prevention of drought damage

Water reservoirs provide a stable water supply by setting the monthly operation level of the dam water before the rainy season (end of June) against the backdrop that severe

Table 6. Flood control effects in 2012.

Main stations		Without dam		With dam		Reduction effect	
		Flood (m^3/sec)	Water level (masl)	Flood (m^3/sec)	Water level (masl)	Flood (m^3/sec)	Water level (masl)
Han River	Yeoju	9,075	35.7	6,305	34.4	2,770	1.3
Nakdong River	Jindong	20,350	14.3	13,230	11.3	7,120	3.0
Geum River	Geumnam	6,374	18.0	1,851	12.9	4,523	5.1
Seomjin River	Gurye	3,304	5.3	1,484	3.7	1,550	1.6

Source: K-water. Note: masl - metres above sea level.

Table 7. Dam flood control effects.

River system	Rainfall event	Precipitation (mm)	Flood control effect (m)
Han	12–17 July	128	3.2
Nakdong	2–8 July	137	2.1
Geum		90	1.2
Seomjin		225	1.2
Yeoungsan		252	No dam

Source: K-water.

droughts occur on average every 20 years and climate change could increase the frequency.

The method of operating the dams' water supply has shifted from determining the discharge amount for each individual dam separately to considering the hydrological situation of the upstream dams and the water demand required when determining discharge volumes. To this end, a real-time water management system was constructed to take into consideration the hydrological situation of the water system, the reservoir water volume, the inflow and outflow status, the power system condition, and water quality and demand. The operating system includes a watershed analysis model, a river water quality prediction model, and an optimal operation model, so that runoff analysis and reservoir operations are carried out systematically.

Flood-related dam operations were initially focused on flood control and securing control capacity and would manage this by emptying the reservoir completely and gradually recovering the water level in the latter half of the rainy season. This approach can have an impact on the quality of life of the inhabitants around the dams, and the ecosystems within the basin areas. Therefore, the focus for dam operations now includes social and environmental considerations to ensure the increase in quality of life and the preservation of ecosystems.

As the rivers, and the factors impacting them (such as green algae), are all tightly connected, an integrated approach is needed to operate the dams as weirs in order to pre-emptively cope with increased drought and the occurrence of green algae in the summer months due to the changing climate.

For this reason, the regulations for dam operation were improved as follows. First, by setting the target level of the dam water to ensure that dam stability and downstream flood control are not affected in the first half of the flood season (at the end of July), it becomes possible to increase the water storage capacity. In the latter half of the rainy season, when flood control is the highest priority, additional water storage can be ensured by setting a stable flood control level (in relation to the flood level limit) through power generation discharge, without discharging flood water. This ensures that each dam is closely connected to other dams, allowing integrated water management in the basin while also ensuring adequate water supply and water quality throughout the year.

In July 2012, there was a severe drought in the central region of Korea, the most severe in 104 years. K-water followed the operating procedure for drought in advance, with the rainfall and runoff forecasting provided by the PFS and RWSS. This enabled the planning of the reservoir operation for the drought and to supply 11 billion m^3 of water from multipurpose dams, 5% more than without this procedure (Figure 13).

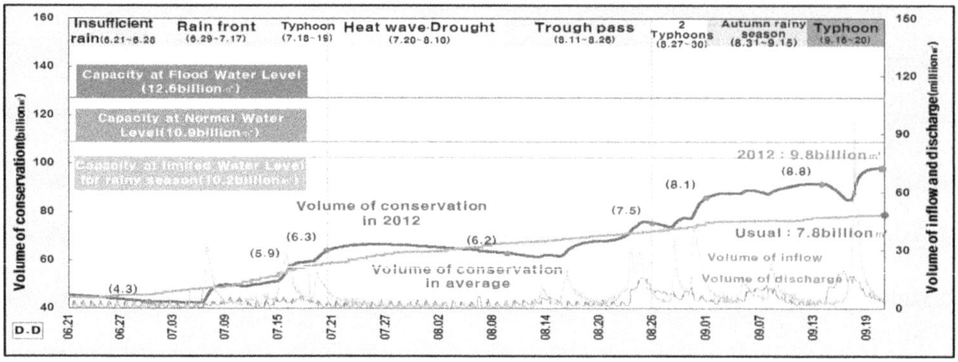

Figure 13. Status of volume conservation in 2012.
Source: K-water.

A total of 8.6 million m^3 water was supplied in advance to drought-prone areas to prevent drought damage, with the flexible operation of the weirs allowing stable intakes despite the severe drought conditions, and 2.7 million m^3 of emergency water was supplied to 29 municipalities by way of waterworks. Moreover, K-water played a key role in water governance by hosting periodic meetings with related organizations and water users. With these efforts, there was no damage despite the severity of the drought.

In 2015, rainfall and water levels of major dams in Korea reached record low levels, in what has since been described as a record-breaking drought. While a 20-year drought inflow has traditionally been used to design the dam storage volume for water supply, in 2015 most inflow into the dams was less than this, officially classifying it as a natural disaster. Continuous drought conditions since 2014 have threatened normal dam operations.

In order to efficiently cope with the drought, water supply adjustment criteria for times of drought were established by K-water. There are four uses of the water from a multipurpose dam. In order of priority they are domestic, industrial, irrigation and instream flow. K-water made an effort to preferentially secure the domestic, industrial and irrigation water by reducing instream flow, as the first three directly relate to the economy and daily life. In order to increase public awareness of the drought, real-time drought situations for each K-water dam are presented on K-water's website.

Despite an unprecedented extreme drought spreading across the country, K-water has been able to stably supply water to the public through the use of comprehensive counter-measures. If K-water had hesitated to carry out these anticipatory actions, interruption of the water supply would have likely occurred. With anticipated and active actions against drought, an additional storage volume of 2.4×10^9 m^3 was secured among nine dams, and drought damage was mitigated.

Improvement of water quality and efficiency

Water quality improvement

Improvement of water quality by way of water discharges first started in Korea in 1991 due to a phenol accident. In March 1991, the Nakdong River Gumi Industrial Complex was flooded, and phenol entered the Nakdong River and then the Dasa Water Supply

Plant in Daegu Metropolitan City. In response, approximately 7 million m^3 of water from Andong Dam was discharged over four days to improve the river water quality. In the case of the phenol spill, Gamcheon, Andong Dam, Imha Dam and Hapcheon Dam were operated in conjunction, with the discharge of 26 million m^3. This was seen again in 2011 and 2012, when approximately 185 million m^3 of water was discharged from the Soyanggang Dam and Chungju Dam to improve water quality in Paldang Lake.

Since the Four Rivers Restoration Project (December 2008 to April 2012), there has been a need for a more sophisticated water quality management response system due to changes in the physical environment of the rivers and the water management environment with the introduction of new river facilities. Public interest in water quality and aquatic ecosystems also continues to grow due to algal blooms and has become a social issue.

In an effort to prevent algal blooms from worsening, K-water opened the gates of the weirs and discharged river water simultaneously or sequentially where the green algae had grown. It should be noted that these operations were carried out by considering the water supply capacity and discharge amount of the dam located upstream.

As it was found that the increase of discharge through dam-linkage operation is effective for water quality improvement, the scientific water management system of K-HIT examined the water quantity of dams and weirs to establish a dam–weir, weir–weir and dam–dam linkage management system to secure an additional 98 million tons of water to be supplied for the improvement of water quality, resulting in a green algae (Chl-a) reduction of up to 25%.

When water is abundant in the rivers, dam and reservoir water is discharged (every 1–5 days) for the purpose of reducing green algae, and the water level of the weirs is maintained (with the water level not affecting the surrounding groundwater), it is predicted that the cyanobacteria will decrease by 22–36% in the Nakdong River and by 21–23% in the Yeoungsan River, and in case of the Nakdong River, the number of days when green algae occur will be reduced to a quarter (from an average of 3.8 to 1), and the number of days exceeding the alert level (more than 10,000 cells/mL) will also decrease slightly (from 51 to 44).[3]

Improvement in management efficiency

By adopting a system such as K-HIT, an increase in efficiency is also gained, as dams and weirs can be operated by fewer people. For example, the hydropower plants are now operated by one-fifth of the staff previously required. This result can be seen across the whole operation of the dams and weirs.

Lessons learned

The role of the public sector is important in implementing Smart Water Management.

Actual, real-time data and new technology are the most basic elements for realizing integrated water management, and it is possible to design short- and medium-term plans based on information on the water management situations in the field. Even though these actual data and new technologies cannot be completed in a short period of time, their importance and development should be recognized through steady investment.

K-HIT is a system for realizing integrated water management, which requires long-term investment and budgeting. While private investment could be a possibility for Smart Water Management (SWM) implementation, the OECD (2011) notes that market mechanisms alone will not provide an appropriate amount of support for eco-innovation, such as SWM technology, at the right time. This is because the private sector would be unable to reap all of the benefits of their innovations, and environmental benefits may not always be appropriately valued. As this is the case, policy interventions and public sector support are required (OECD, 2017).

As the benefits from K-HIT have a public interest, through flood prevention and the reliable supply of water, it is in the interest of the public sector to support these innovations. However, this public interest also limits the private sector's role in the construction of the system, which can be limiting for expanding and replicating SWM technologies.

Step-by-step system construction is more financially efficient and worth considering based on the economic situations of each country.

Since a large amount of money is needed to build an integrated water management system, it is necessary to build a system step by step in order to apply the integrated water management system after each improvement of the existing system in consideration of the economic conditions and budget of the relevant country. K-water has developed a successful approach to constructing the system in several phases. SWM also relies on extensive coverage of ICT and the capacity to maintain and operate the system. A recent OECD report (2017) confirms that while this may be possible in developed countries such as Korea, it may be more difficult in developing countries, where capacity and availability of technology may differ, so when implementing SWM systems in other countries it is important to take into account the social and economic context to ensure its success.

In order to efficiently and effectively cope with drought collaboratively in an integrated approach, it is also important to establish a step-by-step action plan.

Having a shared action plan with easily recognizable criteria for each drought stage provides the relevant stakeholders (including all dam operators) with the tools needed to identify the particular standard water levels for an entire year and also provides a step-by-step water supply adjustment plan according to the drought stage. As a result, the speed of the decision-making process, such as reducing the water supply, has become much faster, and dam operators can now work together collaboratively to ensure a successful, integrated water management approach.

To prevent the drought getting worse, it is necessary to set up measures appropriate to the characteristics of each basin.

Despite the anticipated actions, various countermeasures were carried out as the drought condition expanded across the whole country and the drought situation was getting worse (Figure 14). In the Han River basin, which is the main water source for the Seoul metropolitan area, water from a hydropower dam was substituted for multipurpose dam water due to the gradual drop of the dam's water level. Furthermore, essential water was

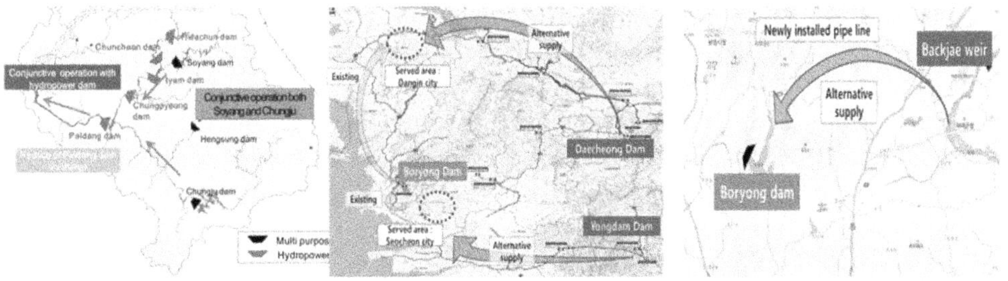

[Joint operation among Multi & Hydropower in Han river basin] [Alternative supply using nearby WSS in Boryeong dam] [Construction of water supply diversion canal]

Figure 14. Additional countermeasures during the severe stage of drought.
Source: K-water.

only provided by joint investigation with the government ministry and relevant local government authorities for actual downstream extraction.

In the Nakdong and Geum River basins, since September 2015, irrigation water has been supplied from dams based on actual demand. In addition, dam water discharges were minimized while maintaining the water level downstream without interruption of water extraction by conjunctive operation among dams, weirs and a barrage. In the Seomjin River basin, a hydropower dam used to produce power generation by diverting water to another area with high head had its release direction changed to the main stream in order to fill the downstream dam storage. In order to prevent interruption of water supply from the dam, various and urgent projects are sometimes required, to prevent the drought from worsening. Despite this, with effective management plans, integrated SWM tools and collaborative stakeholders, these additional measures can be significantly reduced.

The project can effectively accomplish the integrated approach of SWM technologies.

In order to cope with continually changing water management, the flood analysis system was extended from four to five major rivers, including the Yangsan River basin, and the 124 flood analysis points were expanded to 164. In particular, since K-water committed itself to the integrated operation and management of weirs in 2012, river management is more focused on overall river management, rather than dam-oriented operation. In order to improve the accuracy of the water level/flow curves, data were collected with sensors, meters and so on based on the results of the flow survey team. The system has been improved by reflecting on the performance of the Four Major Rivers Restoration Project and on the expansion of its function.

Conclusions

In Korea, two-thirds of the annual rainfall is concentrated in the summer, so floods are frequent, and efficient water supply is essential to protect and supply water to the growing population and to continue economic development. Therefore facilities, including dams and weirs in rivers, and efficient water management have evolved over decades. In recent years, since the frequency of heavy rainfall and drought are increasing due to climate change, the need to effectively manage water resources has increased. It has they become essential to secure the optimal operation of water gates for continuous measurements and analysis of rainfall inflow and outflow, weather conditions and water quality. K-HIT was introduced in 2002, and since then the system has been upgraded continuously to expand the management of rivers and functions.

K-water's flood control capacity accounts for 95% of the gross domestic flood control capacity, or 4.9 billion m^3 a year, and K-water's water supply capacity comprises 66% of the gross domestic water supply capacity, or 12.4 billion m^3 a year. Through the introduction of SWM and countermeasures against floods and droughts, we have been able to effectively deal with floods that occurred in 2012, 2013 and 2015. Consequently, the capacity to operate and manage that system effectively determines Korea's resilience to water risks. SWM can enhance that capacity by collecting and sharing real-time data on water use, expected rainfall, and available room in reservoirs. This application enhances K-water's capacity to deliver its mandate on water quantity management and prevention of scarcity and flood risks (OECD, 2017).

In this way, K-HIT has systematically implemented SWM by introducing ICT to flood management, water supply and hydroelectric power generation, thus improving the efficiency of water management in the watershed by enhancing decision-making capabilities. We have also been able to improve productivity through efficient power generation operations. Finally, we can increase profits by reducing the staff required to manage the system and by improving efficiency through the centralization of hydrological and power-generation controls.

Notes

1. The design flood is the engineering design standard for river facilities considering the characteristics of the flood, frequency of occurrence of flood, potential damage due to flood, and economic factors.
2. CDMA (Code Division Multiple Access) is an example of multiple access, where several transmitters can send information simultaneously over a single communication channel. It is used as the access method in many mobile phone standards.
3. Announcement of the result of the research service 'Operation Plan of Dam and Weir Connection', Ministry of Environment press release.

Disclosure statement

No potential conflict of interest was reported by the authors.

References

MoLIT. (2017). *The 4th long-term comprehensive plan for water resources (2001-2020).*

OECD. (2011). *Better policies to support eco-innovation.*

OECD. (2017). *Enhancing water use efficiency in Korea.* OECD studies on water. https://www.oecd.org/fr/publications/enhancing-water-use-efficiency-in-korea-9789264281707-en.htm

Integrated Smart Water Management of the sanitation system of the Greater Paris region

Jean-Pierre Tabuchi, Béatrice Blanchet and Vincent Rocher

ABSTRACT
This case study details the development of a real-time control system (MAGES) in the Paris region designed to better control stormwater pollution caused by combined sewer overflows and to optimize the need for additional storage or treatment facilities. It is structured to outline the challenges facing the Greater Paris region water and sanitation networks, and the solutions provided by SIAAP, the public utility in charge of the treatment and transport of wastewater, over the past 20 years.

Summary

The greatest challenge the sanitation system of the greater Paris region had to face in the final decades of the twentieth century was quality recovery for the Seine and Marne Rivers. The pollution of the water discharged into the Seine was caused by a lack of treatment capacity and technical performance as well as by overflowing sewerage systems during rain events.

After decades of investments, huge improvements in the water quality of discharged waters were obtained, and the objectives of the European Water Framework Directive (WFD, 2000/60/EC, 23 October 2000) are close to being achieved, thanks to the development of wastewater treatment plants (WWTPs) and a sewage transport system. At the same time, the Syndicat interdépartemental pour l'assainissement de l'agglomération parisienne (SIAAP), the public utility in charge of the transport and treatment of wastewater for the Greater Paris region, has also invested in real-time control following a 1997 study that recommended it for better control of stormwater pollution caused by combined sewer overflows (CSOs), reducing the need for storage facilities.

Building upon existing systems and the experience acquired since the mid-1980s at SIAAP, as well as each of the its constitutive departments (Paris, Hauts-de-Seine, Seine-Saint-Denis and Val-de-Marne), this real-time control system, MAGES (Modèle d'aide à la gestion des effluents du SIAAP), began operation in 2008. The new system integrates all the data from each department and is powered by a hydraulic deterministic model fed in real time by 2000 sensors. It provides flow forecasts for a trend scenario in each part of SIAAP's networks and at each treatment plant on different time scales depending on the weather conditions. This trend scenario is used by the operators to adjust the management of the system.

This smart system takes advantage of the capacity within the coverage area to transfer sewage from one WWTP to another. Such transfers enhance system-wide security in case of shutdown for any reason, including planned works and incidents.

MAGES driven several changes in the operation of the sanitation system. First, each operating site has the knowledge in real time of what happens elsewhere in the system, resulting in a shared and global view of the system. At the same time, the SIAAP department that operates MAGES has a global overview of the hydraulic running condition of the whole system.

Ten years after the commissioning of MAGES, it is still difficult to assess its benefits in terms of savings on either investment or operation costs. Projected constraints on the operation of Paris's regional sanitation system from tighter regulations, population growth and effects of climate change on the Seine hydrology are impelling SIAAP to develop smarter tools aimed at reducing pollutant loads discharged into the rivers without excessive costs.

Background

This section provides an overview of the characteristics of the River Seine (Tabuchi et al., 2016).

Flow rate of the Seine

The median flow rate of the Seine in Paris, and the 5-year and 10-year wet and dry flow rates, are low compared to other French rivers (Figure 1). The Seine and the Marne have an oceanic regime characterized by low flows during summer until the start of autumn and floods in February (due to low evaporation and high rainfall).[1] The flow rates in these two rivers are controlled, for both high and low flow, by upstream storage dams, thus limiting the hazard of natural flooding.

The main characteristic of the Seine flow rate in Paris, and the Marne where it enters Paris, are compiled in Table 1.

Flow control on the Marne and the Seine

Like all hydrological systems, the Seine and the Marne are subject to high-flow and low-flow periods, but the extent of the development of the Paris conurbation has made it particularly vulnerable to these natural phenomena, and to flooding in particular. The floods of 1910 and 1924, but also the drought of 1921, prompted the French government and the local political authorities to adapt the Seine catchment. The final works were completed in 1990.

The Seine Grands Lacs public institution currently manages 850 Mm^3 of storage capacity designed to mitigate flooding, alleviate low flow and help meet the water needs of Paris, as well as the cooling needs of the nuclear power station at Nogent-sur-Seine.[2] As an

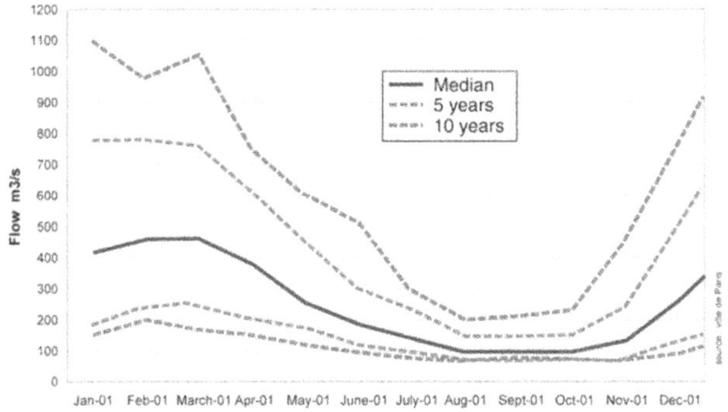

Figure 1. Five-year and ten-year flow rates of the River Seine in Paris, 1974-2010. Source: Ville de Paris/ AESN.

Table 1. Characteristic flow rates of the Seine in Paris and the Marne upstream of Paris.

	Average	Low-water (5-year) average monthly flow rate	Low-water (5-year) daily flow rate over 10 days	High-water (5-year) average daily flow rate	High-water (10-year) average daily flow rate
Seine in Paris (Austerlitz Bridge)	310 m^3/s	82 m^3/s	71 m^3/s	1,400 m^3/s	1,600 m^3/s
Marne in Noisiel	109 m^3/s	32 m^3/s	27 m^3/s	440 m^3/s	500 m^3/s

Table 2. Comparison of minimum water flows and dilution capacity of different French rivers.

	Five-year minimum water flow m^3/s	Conurbation	Population Millions of inhabitants	Impacts Dilution capacity
Rhine	520	Strasbourg	0.7	65 m^3/d/inhabitant
Rhône	380	Lyon	1.8	18 m^3/d/inhabitant
Seine (at Poissy)	170	Paris urban unit	10.5	1.4 m^3/d/inhabitant

illustration of the alleviation of low flow provided by the reservoirs, 40% of the minimum annual flow of approximately 85 m^3/s for the Seine in Paris comes from storage dams.

A megacity on a small river

The Seine, which drains the pollution generated by 14 million inhabitants in the Paris region, has a very low discharge dilution capacity, especially during its seasonal low-flow periods (Figure 1),[3] which is significantly less than other river basins in France (Table 2).

SIAAP and the sanitation at the heart of Paris

SIAAP, created in 1970, is the public utility in charge of the transport and treatment of wastewater for the Greater Paris region. SIAAP is the operator downstream of a large sewage collection and transport system for a drainage area of 1800 km^2 and serving the sewerage needs of 9 million inhabitants. As operator of the main sanitation system on the Seine catchment, it is also responsible for the impact of the sanitation system on the natural environment. Furthermore, like many other sanitation utilities around the world, SIAAP is playing a growing role in a carbon-free and a circular economy.

Table 3. Breakdown of the linear length of the main sewers owned by the four departments.

	Paris	Hauts de Seine	Seine-Saint-Denis	Val-de-Marne
Combined sewers	2,100 km	384 km	356 km	195 km
Stormwater		74 km	190 km	377 km
Wastewater sewers		72 km	124 km	261 km
Total	2,100 km	530 km	670 km	833 km

Source: Tabuchi et al. (2016)

The sanitation scheme for the extended urban area of Paris has evolved over time. In 1929, the principle was to concentrate all of the wastewater at a single downstream plant. In 1968, this single-plant concept was abandoned, at a time that coincided with the emergence of the first institutional decentralization measure for the extended urban areas of Paris.

Sewage collection and transport systems at the heart of Paris

A vast collection and conveyance system, with an estimated length of 15,000 km, has been created over time in line with the growing urbanization of this region (Table 3).

In many cases, the Paris region sanitation is based on a combined sewer system in its centre and on a separate system in its outskirts, which was developed more widely after the Second World War.

Real-time control system

Historically, stormwater management has focused on managing rainwater runoff, particularly to reduce the risk of flooding due to the system overflow. This drove managers to start investing in real-time management at a very early stage. Seine-Saint-Denis, the pioneer, started to implement its first local real-time control station in 1974 for the management of a stormwater storage facility. Ten years later, they had a complete remotely controlled system. Today, each operator possesses a system adapted to its specific constraints, with the dual aim of combating flooding and controlling discharge into the rivers. These real-time control systems are now interconnected, and the different operators communicate with each other.

One of the unique characteristics of SIAAP's system is the interconnection capacities among the WWTPs. This ability to transfer flows among treatment plants is a rare enough occurrence worldwide to be worthy of mention here. The system is managed with a real-time control and decision support system, the SIAAP Sewage Management Assistance Model.

Water challenges

Recovery of water quality in the Seine

The Seine is not a large river when compared to the population settled in its catchment: around 14 million inhabitants rely on a flow as low as 95 m^3/s. This means that one of the key water challenges is maintaining the quality of water discharged into the Seine. During the last five decades, the challenge was the recovery of water quality within the Seine through and downstream of Paris, and to reach the quality target of the Water Framework Directive.

Urban development has led to deterioration in the quality of the Seine's water since the 1870s and until the 1970s. During the last 35 years, the physico-chemical quality of the water in the Marne and Seine has improved significantly, both upstream and downstream of the extended urban areas of Paris.

Growing population in the Paris megacity

Throughout the SIAAP's administrative territory, a 9% population increase is anticipated from 2012 to 2030,[4] which amounts to an increase of a million inhabitants, which will significantly increase the water management challenge: water supply, wastewater collection and management of the impervious surfaces of new development areas.

The effects of climate change

The effects of climate change on the hydrology of the Seine have been examined by several research projects, in partnership with operational stakeholders. All these projects concur that major changes will take place in the hydrology of the Seine basin area from 2050 onwards: water and resources will be increasingly scarce, and flood hazard will persist.

The main conclusions are:

- An increase in air temperatures of 2.3 °C for the annual average temperature, and as much as 3 °C in summer.
- A downward trend in summer rainfall, with strong uncertainty about the rainfall characteristics, but of lesser importance than the increase in the evaporation rate.
- The rising air temperature will significantly increase the potential evapotranspiration demand. At the scale of the Seine catchment, this will strongly reduce aquifer recharge and lower the flow rates in the rivers.
- The low flow rates of the Seine will drop significantly. A drop of 10–50% in the mean annual flow rate of the Seine in Paris is anticipated, depending on the models used. The drop in the five-year low flow rate could be as much as 60%. This decrease is simulated on the basis of the current abstractions and land uses. The changes in 10-year floods are less consistent and are generally statistically less significant.
- In aquifers, the groundwater table will sink by 1–5 m compared to today's level, impacting the main sources of supply for the rivers.

The analysis finds that by 2050, the greatest risk is drought. The main issue relating to climate change concerns the risks of deterioration in water quality as a result of the drop in the flow rates of rivers. Overall, the pollution loads discharged by the urban WWTPs in the upstream basin areas of the Seine are likely to remain stable, as the population will not increase as much as in the Paris region.[5] Any drop in the flow rate should lead to higher concentrations of pollutants in the water. Paris will thus face a twofold problem, with less dilution capacity and a highly likely increase in pollution concentrations in the Seine, Marne and Oise. At the same time, the growing population will increase the pressure on treatment systems. In this context, if one assumes that sustainable development is based on maintaining good water quality for the next generation, maintaining the water quality within the Seine becomes a very important issue.

Smart Water Management solution

MAGES presentation: how does it address the Paris region's water challenge?

The development and implementation of the Smart Water Management systems in the Paris region is the result of a long process originally related to local floods during cloudburst events. Since that time, the sanitation system has gone through major changes in its structure and its regulatory frame (Blanchet et al., 2008; Tabuchi & Blanchet, 2016).

Today, sanitation in the Paris region relies on a multi-stakeholder management organization and a vast sewer system to collect and transport the 2.4 Mm^3 per day produced by 9 million inhabitants to six interconnected WWTPs, and all the systems have a high risk of impacting the quality of river waters. Following the development and implementation of a local real-time control system, a comprehensive vision of sanitation system operation was developed. This appeared to be increasingly necessary to efficiently manage this interactive and complex system to comply with Water Framework Directive objectives.

In 2001 SIAAP decided to build a new real-time control system on the base of the existing systems. This new system had to allow real-time data and information exchanges between the different systems, to provide a comprehensive overview of the status of the system and to produce operation scenarios. This system is based on hydraulic models fed by continuous monitoring of the sewage network and weather-forecast-based radar images. We now provide a description of the solution and the main results of implementing these models.

MAGES context and origins

The technical context

The network under SIAAP's control consists of large collectors and main sewers, built at depths from 3 m to 100 m, depending on the topography, and with diameters from 2.5 m to 6 m. Its total length is about 400 km. Numerous system interconnections have been created to guide flows according to the available capacity of downstream main sewers and storage and treatment works, and also to allow bypass without spillage of untreated water to receiving water during maintenance. One of the very special characteristics of Paris is its high capacity to transfer water between WWTPs. Although it is common in large cities to find several WWTPs, it is rare that they have the possibility of transferring water between them through an interconnected sewer network. This feature is the fruit of the history of the agglomeration. Originally the system was designed with a single WWTP downstream of Paris. In 1968, a sanitation master plan was adopted with the idea of splitting the collection zone and distributing treatment facilities around the city. This distribution system was strengthened in the 1997's sanitation master plan, whose main conclusion was the decision to downsize the Seine-Aval WWTP's capacity from 2.7 Mm^3 to 1.3 Mm^3 per day. This evolution from a single treatment site to a multi-site approach based on a main transport network has paid off, and it has been taken advantage of by reinforcing this link between the treatment plants. The flow management system takes great advantage of this feature.

As a consequence of downsizing the Seine-Aval WWTP, its water velocity decreased, which, with the small slopes in the network, resulted in an increase in solid deposits in the sewage transport network. This is a problem during heavy rain events after a long dry period: the consequence is a very high suspended solids concentration.

The other important aspect is the water management during rainy periods. In order to reduce the CSO frequency, volumes and impact on the receiving environment, 955,000 m^3 of storage tanks and reservoir tunnels were built to ensure the storage of the polluted water and its release to the treatment plants.

Approximately 150 local management stations with remote-controlled regulating and pumping equipment have been deployed in the SIAAP's area of competence, enabling real-time control of the works.

Among the important technological developments, the treatment processes implemented at SIAAP rely heavily on biofiltration, which is distinguished by the short residence time of the effluents (less than 2 h). This means that the treatment systems are very responsive, with very little buffer time. This is an advantage, for example, in adapting the wastewater treatment required to protect the receiving waters; but it is also sometimes a disadvantage: the operation of such a system can be difficult due to its reactivity.

The road from SCORE to MAGES

In the late 1980s, SIAAP developed a real-time control system, called SCORE (Systéme de coordination, d'organisation et de régulation pour l'exploitation des émissaires), to address the significant shortage of effluent treatment capabilities. At that time, only three WWTPs were operational, the capacity deficit was around 500,000 m^3 per day, and they only concerned the 'Western system[6] 'related to the Seine-Aval plant, whose project was then to increase its capacity from 2.1 Mm^3 to 2.7 Mm^3 per day. In an interim stage, the idea was to take advantage of the sewer pipes' ability, in dry weather, to store the daily peak of the effluents and to treat the sewage during the night, when the flow is lower. However, flood risks during rain events need to be managed. Figure 2 and 3 show the results of flow control to smooth the hydrograph between night and day. Figure 2 illustrates the regulation obtained with SCORE at its beginning, when the average flow capacity of the Seine-Aval WWTP was around 25 m^3/s, and the improvement of flow control from year to year.

SCORE was based on the system of five main sewers supplying Seine-Aval, which were equipped with regulating valves. Each main sewer was divided into several trunks used for sewage storage. Given the hydraulic risks associated with the rains, this system already included a rain forecast. The remote management system was the precursor of the current

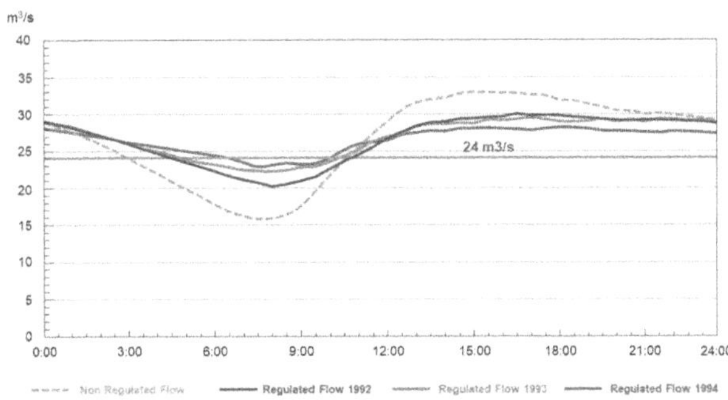

Figure 2. Illustration of flow control at the inlet of the Seine-Aval wastewater treatment plant with the SCORE system. The curves represent the average daily flow variation without control (dashed curve), and the improvement of the regulation obtained between 1992 and 1994 (other curves).

system and began operation in 1992. It was independent of those implemented by each department.

Figure 3 shows the current flow control with MAGES in dry conditions. The steady flow is required by the operator of the Seine-Aval WWTP.

In 1989, following local protests, it was decided not to increase the capacity of Seine-Aval as originally planned. This decision led to the development of a new sanitation master plan, which was approved in 1997. This scheme introduced three important changes:

- Downsizing Seine-Aval's capacity to 1.5 Mm^3 instead of 2.7 Mm^3 per day. This led to a new distribution of sewage treatment capacities on SIAAP's territory.
- Stormwater pollution control due to CSOs. The master plan was designed to avoid untreated CSO spillage for a return period of less than 6 months. With this in view, the construction of a total stormwater storage capacity of 1,693,000 m^3 was planned.
- Implementation of a real-time control system for SIAAP effluents. The studies of the time showed that a real-time control system would allow the reduction of 500,000 m^3 in discharges from the sanitation system during heavy rains.

MAGES's objectives

In this context, in 2001, SIAAP, with the active collaboration of all its partners, launched the construction of a dynamic management support tool for its sanitation system, MAGES. This system had three main objectives:

- Global and shared knowledge of the sanitation system. To satisfy this objective, the operators pool the data they use for the management of their works through a data exchange system: EDEN (Environmental Data Exchange). EDEN allows the exchange and centralization of the measurements and the current state of the works.
- The prediction of the operation of the sanitation system in a stable configuration. The aim is to provide each operator with a forecast of the flow capacity of the sewage transport system and storage works and the flow arriving at each WWTP, with a forecasting horizon of 24 hours in dry weather and 6 hours in rainy weather.
- Assistance in the management of the sanitation system to optimize the use of the structures during the rainy episodes.

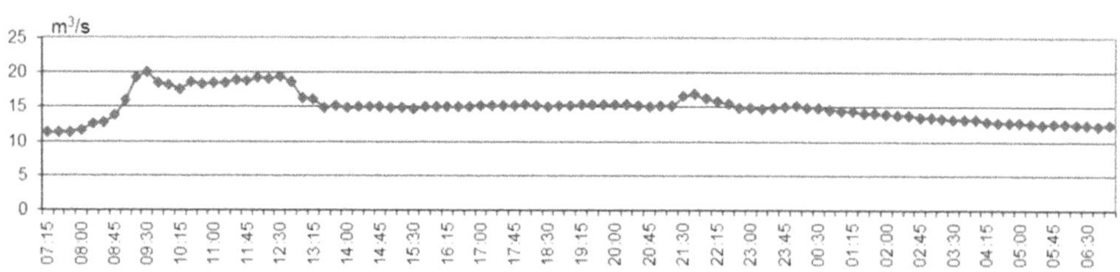

Figure 3. Example of remote flow control at the Seine-Aval inlet with MAGES in dry-weather conditions (2 July 2018).

The development process

On this basis, SIAAP launched SCORE's evolution project towards a system integrating all the SIAAP networks by opening and extending it to equivalent systems of its partners.

It began with studies and interdepartmental working groups (with members from SIAAP, departments and the Water Agency) aimed at pooling knowledge about the structure and operation of the network. This stage was the occasion for establishing in-depth and intense exchanges between the stakeholders. The studies were followed by developments of the existing SCORE system. This system allowed the supervision of the flow conditions (water level and speed) in the entire SIAAP network, as well as remote control of the control devices. In addition, EDEN has been developed to enable SIAPP and network supervision centres in the core zone to share data in real time. At this stage, SIAAP had knowledge of the real-time structure and operation of its network and upstream departmental networks. It had tools to control the network devices, so flow management could be implemented operationally from remote centres. However, given the complexity of the network, a management support system was essential.

Given the innovative nature of the management assistance tool to be put in place, the uncertainties in the results, and taking into account the obligation to achieve the goals stated, the second step was to launch a kind of contest. According to the French Public Procurement Code, two *marchés de définition* (definition studies) with two consultant consortia were selected in January 2004.

The study was divided into three phases: modelling, decision support tool, and tender document. The objective of the modelling was to develop a software model to reproduce and then predict, using measurement data (hydraulic, rainfall) and other information (positions of moving parts, availability of works, etc.), factors that can be used for the management of structures.

The decision support tool relied on the modelling effort and the methods proposed by the two consultants. The main result of this phase was a software model which, fed by the same type of data as the first model, was to provide management instructions that could be used by the operators of departmental and interdepartmental sanitation works.

The tender document[7] specified the detailed requirements for the creation and full implementation of the MAGES forecasting and decision support tool.

These studies began in April 2004 and ended in October 2005 with a tender document and the technical and financial offer for the development of the tool. Two approaches were proposed: a solution based on stochastic modelling, and one based on deterministic modelling of all the processes from runoff to hydraulic transport. The deterministic model was chosen because it was considered by the operating teams more understandable and manageable in its adaptations and evolutions. The realization of this project was led by the consortium constituted by the companies Eau et Force (a SUEZ subsidiary) and SATELEC.

This led to the launch of the MAGES project, which is the core of the real-time control system. The aim was to put a forecast model of the network states and a management aid system at the disposal of the operators of SIAAP and the departments of the near suburbs of Paris. This system provides instructions to the control devices in order to reduce spills in the Seine and Marne, and to optimize treatment.

The technical solution

General principles

MAGES relies on real-time deterministic modelling of transport and storage structures with 23,000 calculation points coupled with an optimization process for searching management scenarios. This modelling considers the following inputs based on a data-sharing network (Figures 4 and 5):

- dry weather sewage flow inputs,
- measured and forecast rainfall inputs,
- field data based on more than 2,000 sensors, and
- physical changes to the sanitation system; these could be related to maintenance works or incidents inherent in the operation of any sanitation system.

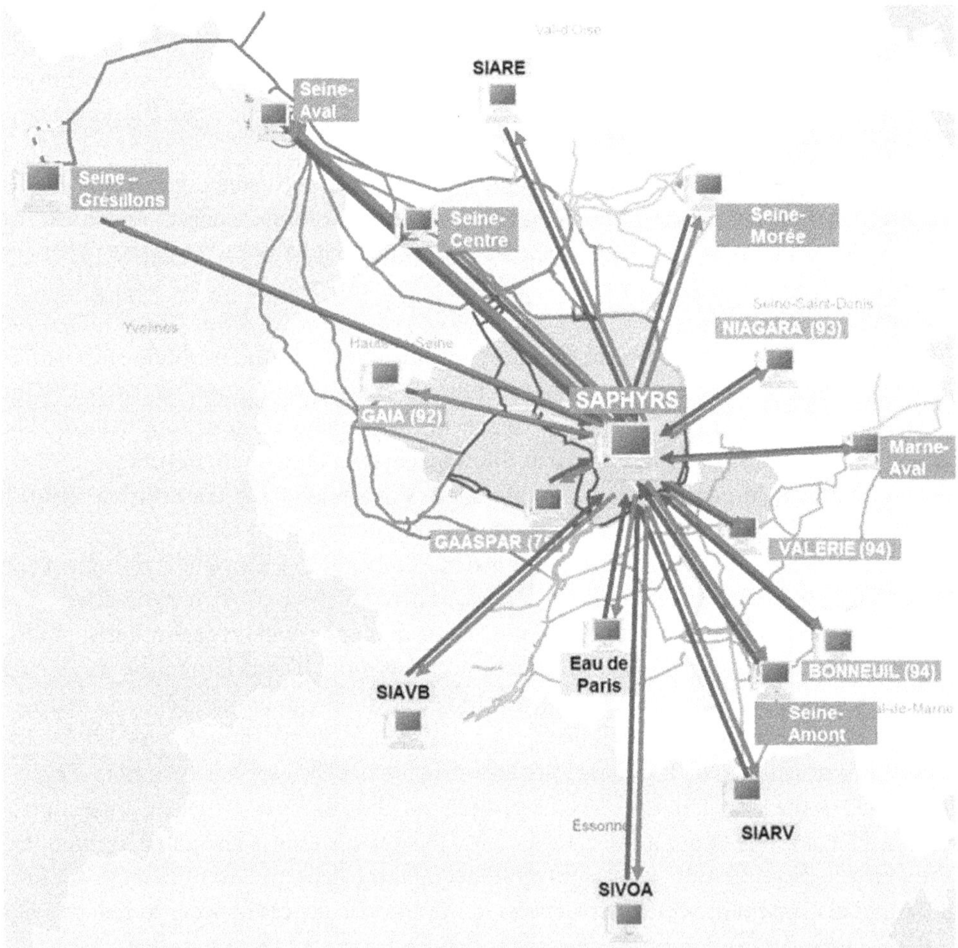

Figure 4. Main data transfers between SAPHYR control room and MAGES, with wastewater treatment plants (green), SIAAP operator members (blue) and outside operators (pink).

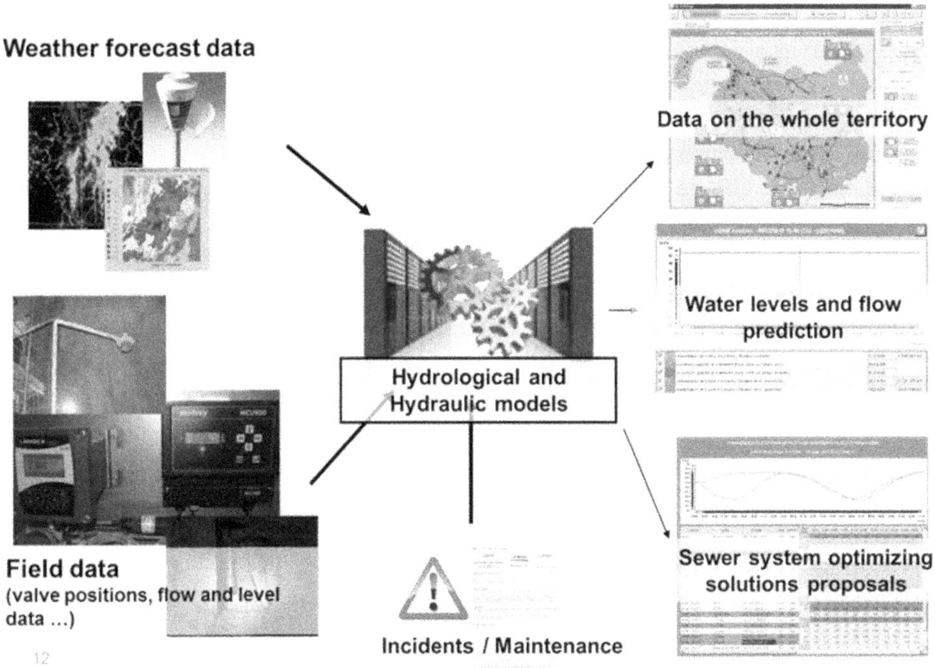

Figure 5. Schematic of data and MAGES output processing operations.

A MAGES treatment cycle updates every five minutes to show:

(1) *Identification of the current state of the system*:
  (a) the measurements collected in real time (after validation);
  (b) these measurements are also used to update the models (position of regulators, filling of basins, etc.); and
  (c) the simulation of the state of the network carried out with the detailed hydraulic model. This simulation provides a consistent state of the system by completing the valid measurements with the results from the modelling. Thus, each operator has at the same time a synthetic, comprehensive and complete vision configuration and hydraulic status of the sewerage network, e.g., heights, flow rates, discharged flows, volumes stored and volumes treated.

(2) *Prediction of future trends*

The MAGES system provides the expected state of the system by integrating the effect of the rain forecasts. It is based on the detailed hydraulic model: flow rates and water levels are estimated assuming that the network configuration (valve position, pumping rates) is unchanged over the prediction horizon, except for works which have a local regulation, which is taken into account in the forecast. The hydraulic status (flows and heights) is then displayed for around 400 key points along the network, with a horizon of 24 hours in dry weather and 6 hours in rainy weather.

(3) *Determination of the optimized management scenario*

A linear model based on simplified hydraulic and hydrological equations calculates a large number of simulations in a few minutes, the results of which contribute to the creation of an optimization problem. This optimization problem is given to a solver that minimizes the cost of an objective function. It is the parameters of this objective function that set the priorities such as limiting spills to the different weirs, the priority of use of the different available storage volumes, or the operating time in the rainy season configuration of the treatment plants.

The simplified model is recalibrated with the detailed model at each 15-minute computation cycle. This solution ensures rapid calculations and convergence of the optimization problem while ensuring realistic and stable solutions, whatever the weather conditions. The scenario is presented in the form of objective flows calculated by the simplified model for 26 key points along the network, selected for their management potential. These flow objectives are also translated in the form of instructions for the 50 or so main control structures (valves, pumping groups, etc.) of the sanitation system. They are presented to the operators as valve positions, flow, or water level instructions, depending on the management mode of the structure.

(4) *Prediction of the optimal trend situation.* The instructions for the main structures obtained in the previous step are injected into the detailed hydraulic model to provide the state of the system in the event that the proposed instructions are actually applied.

There are three objective functions, corresponding to the three management strategies that have been set in MAGES:

- The long-term dry-weather strategy aims to smooth overall contributions to WWTP while not increasing the risk of spills.
- The overflow rain conditions strategy aims to minimize spills without increasing the risk of floods by overflow during medium-to-heavy rains. By anticipating the emptying of the collectors, it optimizes the treatment at the WWTP and the storage capacities in the basin and network.
- The rain overflow time strategy is designed for heavy rains, primarily to reduce the risk of overflow of the network onto the roadway. All the outlets are then used to relieve the network.

Even if there are automated local control devices, the operator is able to act directly on 160 remote devices, such as valves and pumping stations (Figure 6).

Current state and trend prediction by detailed modelling

MAGES relies on deterministic modelling of the urban water cycle,[8] from the hydrology of the perimeter (transformation of rain heights into runoff to the network) to the hydraulics of the structuring network of the sanitation system. Accurate real-time reproduction by a detailed model of dry and rainy weather inputs, network configuration, and meteorology conditions feeds into the results of the modelling. Taking them into account is a real technical challenge, whose main characteristics are presented below.

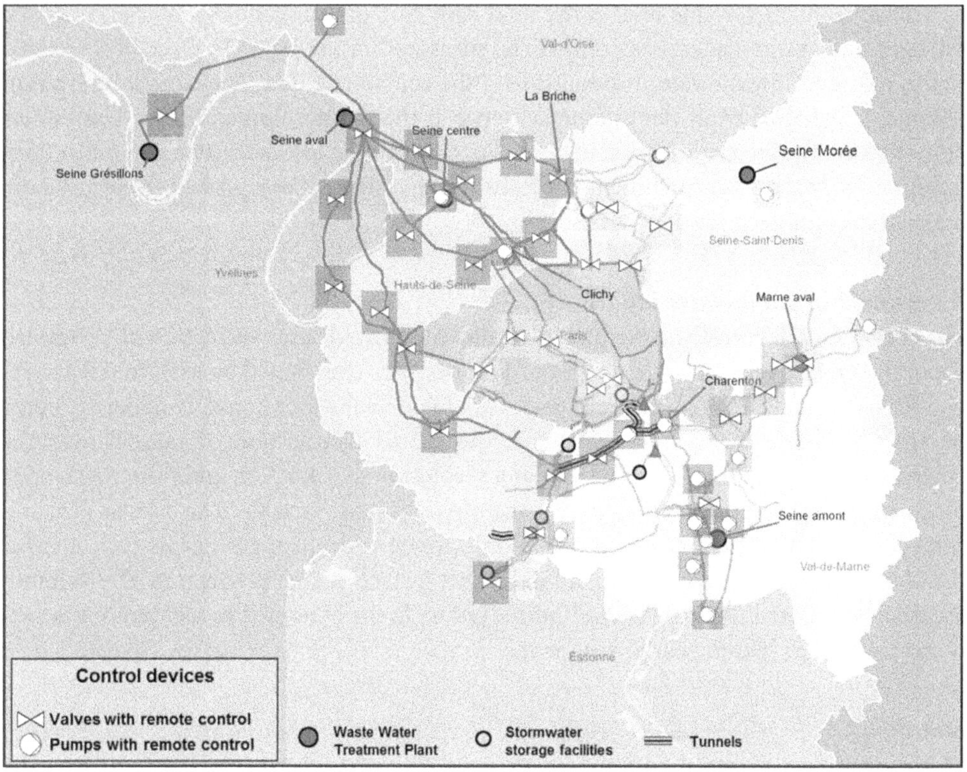

Figure 6. Main remote control devices. Colour code for sewers refers to the operator: red for SIAAP, grey for Paris, orange for Seine-Saint-Denis, green for Val-de-Marne.

Dry weather flow assessment

The MAGES perimeter was divided into 60 input sectors, which are broken down throughout the network by 153 sewer inflow points. In order to reproduce the weekly and monthly fluctuations in wastewater and the seasonal variations in permanent infiltration water inflows, which have been shown to correlate directly with groundwater levels, a process of recalibration of dry-weather flow has been developed in MAGES.

For each catchment area, feed hydrographs of wastewater and infiltration water were constructed from the archived data. For the real-time adjustment of these hydrographs, the simplifying hypothesis is that the global volume of dry-weather inputs modelled for the current day is best approximated by the volume measured at the treatment plants for the last known day of dry weather. A statistical analysis of three years of self-monitoring data of the WWTP showed that the total volume of inputs calculated from the measurement of the last known day's contribution is a good estimate, within a confidence interval of ±11%, for 95% of cases. This method thus ensures that changes in dry-weather production are taken into account correctly, in particular seasonal variations in infiltration water and wastewater inputs (especially during the summer holidays), as well as interannual changes in wastewater production, particularly related to changes in consumption of drinking and non-drinking water.

The adjustment variable used is the total daily flux of nitrogenous materials (organic nitrogen and ammonia, expressed in NTK) measured in the WWTP, directly correlated to the volumes of wastewater produced over the catchment area. The overall infiltration volume is deducted from the difference between the input volume and the wastewater volume. Finally, the correlations between the total volume of wastewater and infiltration water and the values per catchment of these two quantities make it possible to adjust daily hydrographs for each input basin.

Assessment of stormwater inputs

For accurate cartography of the precipitation data, SIAAP uses the CALAMAR model. From the meteorological radar images of Trappes (78), distributed by METEOFRANCE, and measurements from the 80 rain gauges operated by the four constituent departments of SIAAP, CALAMAR provides data on the spatial distribution of rainfall over the covered area at 5-minute intervals and at a resolution of 1 km^2. In addition, the 2-hour forecast, at 5-minute intervals, is used as the MAGES input variable. The data on past and projected precipitation are then mapped onto the territory using radar images, divided into 1200 watersheds, representing a total active surface of 28,000 hectares. The volumes are then distributed to the 388 sewer inflow points in the hydraulic model (among which there are 153 dry-weather inflow points).

Metrology

Metrology is an essential component of MAGES: it builds an understanding of the current state of the sanitation system and updates the models used for the elaboration of the trend and optimized situation. Metrology is a set of more than 2000 variables corresponding to flow, level, valve position, or flow rate measurements. A measurement validation process based on several complementary methods was set up on each of the measurement points to ensure the quality of the input data of the system. In addition to these systematic and automatic validation methods, authorized operators have the ability to intervene to inhibit the measurement of a given sensor, or even correct the value if they know it.

The detailed modelling

The modelling tools are the METE-EAU software package for the hydrology of the upstream zones of the network and HYDRANET for the hydraulic modelling, both developed by Hydratec. These tools have been adapted to meet the constraints of fast execution of calculation cycles and robustness of operation. Indeed, the modelling extends over 500 km of main sewers, modelled by 3113 user nodes (and translated into 23,000 computation nodes); 150 management stations (basins, valves and pumps) were modelled, integrating the automated local management rules for existing ones.

To satisfy the execution time constraint (approximately 3 minutes for a 24-hour simulation), several processes for optimizing the calculation time have been implemented, notably the 'hot restart' of the results of the previous calculation cycle and a division into five interconnected sub-models. Calculations are performed sequentially by sub-models according to an upstream–downstream logical tree (Figure 7).

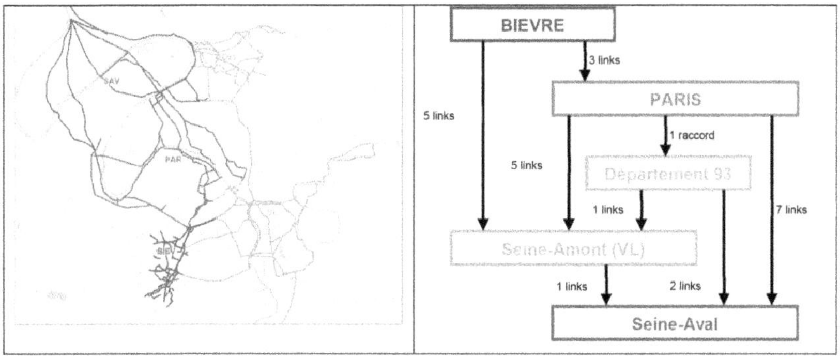

Figure 7. The five hydraulic sub-models and their links.

The MAGES update process

In order to guarantee the coherence and reliability of the results calculated by MAGES compared to the field reality, the model is updated on a 5-minute cycle for all the information influencing the result of the predictive modelling. During this process, the following inputs are taken into account: the measured and predicted rainfall, and the measurements (positions, instructions, flows) and configuration of the works which result from the scheduled maintenance, but also from the interventions and incidents inherent in the operation of any sanitation system.

A first update function is performed on the control and storage structures, at each 5-minute cycle, using the previously validated measurements.

A second function of updating has been developed to take into account the maintenance works. These works are programmed several months in advance and are communicated to the model via a specific screen interface. In the same way, information regarding flow diversion (to avoid raw water discharges) or in the management rules of a control device is taken into account. Once the operator has validated the beginning of the work, the model automatically updates itself on these new configurations of works, as of the following simulation cycle.

This update is also applied for any incident that occurs on the network, such as a set of manoeuvres routing effluent following a WWTP operating problem.

Finally, in the absence of the updated data, the model reproduces an operating dynamic management based on the rules that are routinely applied by the operator for each local management station. The result is a baseline scenario, which provides consistent and credible results in a fallback situation, that is, when none of the update or registration data normally provided by metrology is available.

The optimized scenario

The optimization aspect of MAGES includes an optimizer, which minimizes the cost of a multi-objective function and a simplified (quasilinear) model used to assess the cost of this multi-objective function in different configurations of the main devices for the network management.

The multi-objective function reflects the cost of spills, overflows, storage in the works, and the use of the stormwater pollution treatment systems. The room for manoeuvring that this function allows is in the saturation of the dry and rainy-weather treatment capacity of the WWTP, the filling of the storage facilities, the use of the interconnected sewage network, and of course the use of different CSO devices.

The optimizer used is the MINOS solver developed by Stanford University. The simplified modelling and control of this optimization are carried out with CSoft software, developed and distributed by BPR-CSO. It is a model of the input-output type, able to calculate the flow, to reproduce the flow rate in the pipes and to explicitly satisfy the continuity equation at the nodes. The parameters of this model are calibrated from the validated measures where they are available, supplemented by the results of the detailed model for the other points (in particular those which are not instrumented).

The optimization module builds the optimization problem for the 50 key piloting devices that have been selected. The result consists of the optimal instructions that are proposed for each of these control devices, over a control horizon of 6 or 24 hours.

The automatic construction module of the problem formats the equations and inequality factors, which need to be optimized to solve the problem. The equations reflect the respect of material balances at the management stations and the hydraulic constraints (i.e., flow rates to be respected). The inequality factors represent the intervals to be respected (minimum and maximum flow at the WWTP). This module uses the results of online calibration to construct dynamic flow equations. This same module then establishes the multi-objective function using the weights and penalties that are specific to the strategy in use.

In order to compare the gains of the optimized scenario with the trend situation on the basis of the same detailed simulation tool, all the instructions proposed for the optimized scenario are incorporated into the detailed model. Thus, the operator is able to judge the impact and relevance of these instructions, comparing their effects to the current management. The calculation resulting from the application of these instructions corresponds to the optimized trend scenario.

The non-real-time mode

Besides the real-time system, there is a non-real-time mode. This mode allows several possibilities:

- to work on past events in order to study these events and learn from experience;
- to make studies for maintenance shutdown – this allows managers to study the best scenario to avoid spills, or to reduce them when there is no other option;
- to test different operation scenarios; and
- for training sessions.

This non-real-time mode is also necessary to replay past situations, because the system keeps all the real-time results from the last three months in its memory. Beyond this limit, the non-real-time mode is used to ‘replay’ the situation with all those data, which are all stored in the database.

MAGES, an operations tool

Simplified architecture and processing cycle

The architecture of the system is based on the principle of modularity (Figure 8). The detailed model (METE-EAU and HYDRANET), the optimizer (CSoft with MINOS), the human–machine interface, the database where measurements and results are stored, the reception of meteorological data, and metrology constitute many independent modules that interact with each other through the supervisor (controller) system.

The supervisor (controller) system was developed as part of the MAGES project, with the user interface. The supervisor sequences the exchanges between all the other modules. Its operation stems from the explanations provided previously:

- On a time step of 5 minutes, it retrieves and filters metrology and meteorology data, then updates the detailed model to calculate the current situation.
- Each quarter hour, it successively launches the detailed predictive model, the simplified model that produces the optimal set points, and then again the detailed predictive model. This results in displays of:
  - o the reference trend situation,
  - o the optimum management scenario, and
  - o the trend optimized situation.

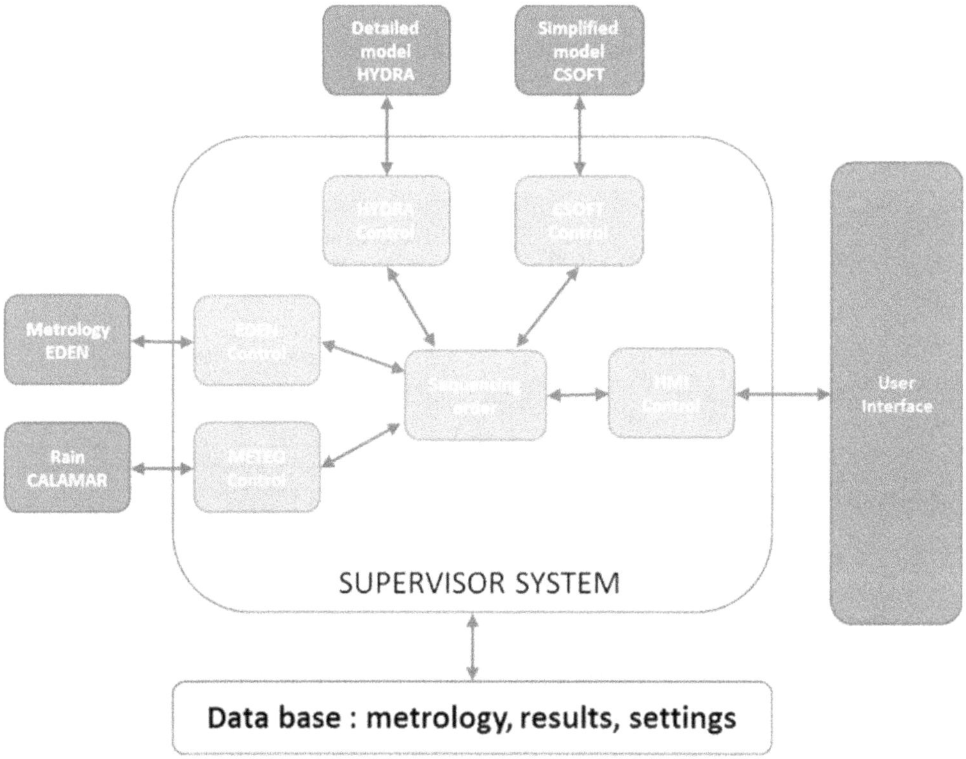

Figure 8. MAGES: general computer architecture.

Figure 9 summarizes the nominal processing cycle of MAGES.

The hardware and software IT infrastructure are implemented redundantly to ensure high availability.

Beyond the various applications needed for configuration that have been delivered to the administrators of MAGES, two main applications are intended for users:

- The human–machine interface for decision–making application, which provides access to all the features mentioned in this article (current situation, baseline trend scenario and optimized trend scenario); and
- The Maintenance Shutdown application, which provides information on planned shutdowns.

MAGES is accessible under the same conditions from the dedicated computers installed in the operators' control rooms, and from the computers inside the SIAAP computer network. An Internet connection also allows the duty staff to access the same information from outside their office.

A global and shared vision of the system

The coherence of all the data provides a global and shared vision of the operation of the sanitation system. It places operators in a global management framework by providing them with information that goes beyond their strict management scope. With the current situation, upstream operators can know the impact of their actions on the downstream network.

Figure 10 is a screenshot showing the entire perimeter of MAGES. It includes the structural collectors, the six WWTPs and their operation setting (dry or rainy weather), and the piloting stations that are receiving instructions. Thematic indicators are also

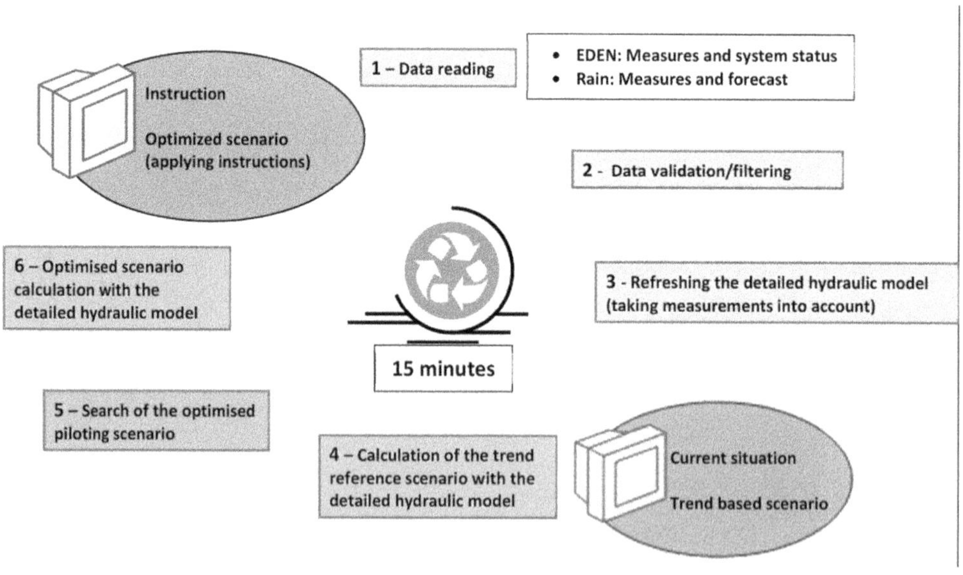

Figure 9. MAGES: simplified processing diagram.

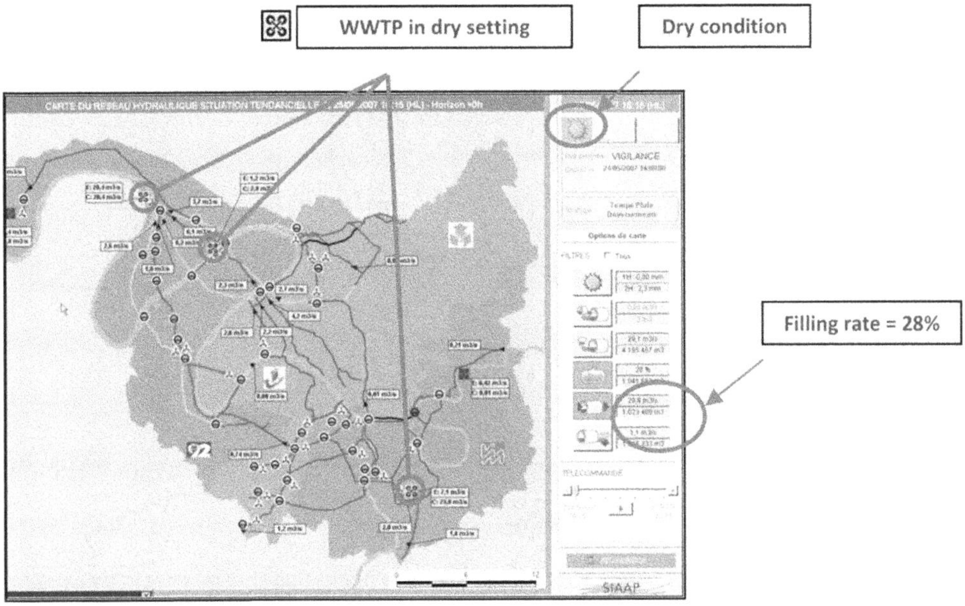

Figure 10. Global overview of the system and thematic indicators.

available, such as actual rain heights and rain forecasts, dry and rainy-weather hydrological conditions, the stored volumes and the storage rates in the different sewers and storage facilities and the flows rates. Each of these indicators is presented in the form of a curve (Figure 11) calculated on the past compared with observations (which can vary from 0 to 72 h) and a forecast on the control horizon, which is 6 h in rainy conditions or 24 h in dry conditions.

This general vision can be zoomed in on each of five major input areas (Figure 12). Each control station has a synoptic window that describes its overall configuration using symbols common to all operators. These synoptic windows also constitute a portal for access to local information.

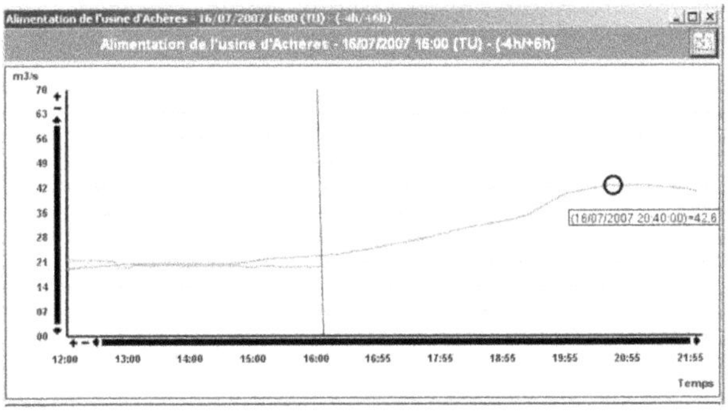

Figure 11. Forecast at 16:00 of a peak flow of 42.6 m^3/s at the Seine-Aval wastewater treatment plant at 20:40 (green curve: measurement, blue curve: modelling).

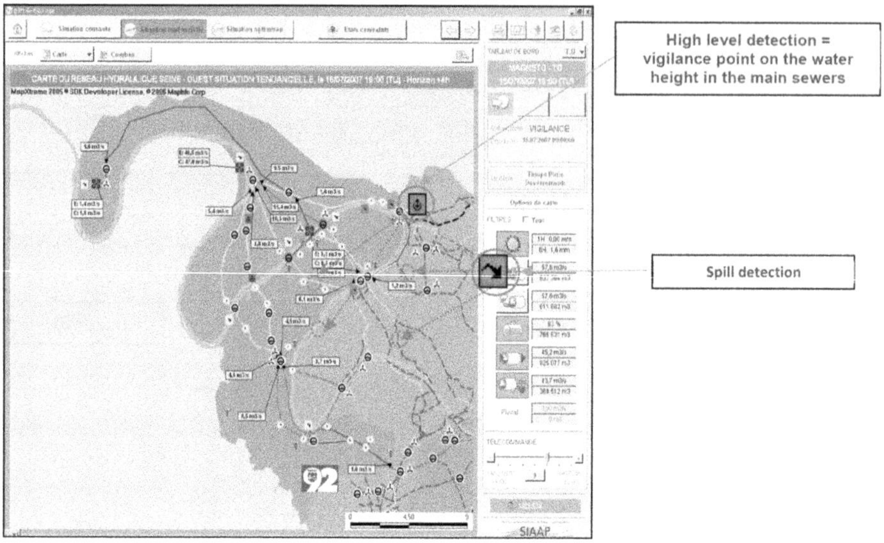

Figure 12. At 1600 h, high levels anticipation of 4 h, dumping points and storage rates in the western network, scheduled for 2000 h.

The forecast and the optimization

The trend situation for operators is predicted by the forecast of very high levels of water or flow in the network (Figure 12), the prediction of the location of spills, the filling rates in the sewers and the operation during rainy weather of the treatment plants over a horizon that varies between 2 and 6 hours.

All information in the previous figure is also available in the form of predictive curves for the various important locations of the sanitation system, such as the flows entering the WWTP, entering and exiting the storage basins, arriving at the inter-connection points, on the CSOs, etc. (Figure 13).

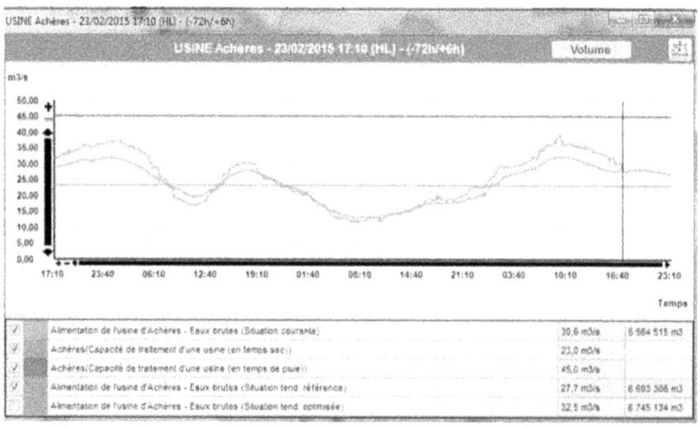

Figure 13. Screenshot of flow prediction at Seine-Aval wastewater treatment plant (green curve: measurement, blue curve: model).

Conclusion

The technical solutions implemented in the construction of MAGES made it possible to satisfy the objective of global and shared knowledge of the sanitation systems of the Paris region for different actors. This technological advance will be accompanied by profound changes in the operating methods of water treatment plants and networks. Indeed, with MAGES, the flow management in the Paris region is based on weather conditions and operating constraints, and technicians are then able to make responsive and concerted decisions. In addition, the decision criteria for the management of the network are built on the scale of the region, no longer only at the local level.

Beyond its real-time use, MAGES is also a tool for studying and assisting the design and operation of new works, in order to make the most of available processing capabilities by optimizing the volume of existing or projected storage facilities.

Achievement and impacts: the contributions of MAGES and smart management

MAGES has brought a radically new vision through remote management, assisted by powerful processing capacity and interconnected network, allowing modification of the distribution of flows between WWTPs with much more flexibility and a global vision. MAGES greatly facilitated the assimilation of the changes: transitioning from static and local, management has become dynamic and global. The vision of the sanitation system has been radically changed.

Various forecast horizons

One of the peculiarities of MAGES is that it allows vision and use at different time scales:

- Twenty-four hours is the current real-time operation, with the sharing of data, instructions to optimize the solicitation of transport, storage and treatment works, taking into account the recent past of dry weather or not.
- One to ten days is the current operating horizon, with detailed and optimized programming of maintenance operations, resulting in a provisional announcement of the status of usage.
- One month to a few months is the horizon of the maintenance works operation. It allows the fine adjustment with all the last updated data for good modelling conditions.
- One to three years is the time scale of the multi-year programming of shutdown operations and specific studies for particular operations.

Related to these time scales, there are several outputs for sanitation system operation. The following sections detail some of these uses.

A shared and global vision

The first objective of providing a global and shared vision has been achieved since the commissioning of MAGES with the consistency of all the data that provides an overall vision of the operation of the sanitation system. It places operators in a global management framework by providing them with information that goes beyond their strict management scope. The shared vision made it possible to set up real cooperation between the different actors, which is part of a process of continuous improvement of the operation.

This shared vision materialized concretely, but the effects of transfers of water with sometimes very different qualities accelerated the process of sharing information. Today, bimonthly meetings bringing together plant and network operators ensure the analysis of past situations, the optimization of operations, the sharing of experiences, and thus a process of continuous improvement of operations, based on the data provided by the MAGES tool. These exchanges have become essential to the proper functioning of the sanitation system.

Optimizing real-time operation

Based on the prediction of dry and wet-weather inputs, MAGES provides in real time, at all points in the network, the flows and water heights at the entrance of the WWTPs and at the characteristic points of the transport network.

The real-time knowledge of hydrographs with an optimized forecast over the next 6 hours provides relevant and essential information for operators to make the necessary decisions on the conduct of their installations. This predictive management makes it possible to better anticipate (1) the start-up of the specific stormwater pollution treatment facilities or to change WWTP configuration from dry-weather to rainy conditions; (2) the regulation of the flow rate on the treatment units to avoid saturating them with a peak of pollutant load likely to exceed the design capacity of the treatment units; and (3) the coordination of the phases of storage and emptying of different reservoirs and tunnels tanks.

Forecast bulletins for the operation of networks and WWTPs

Faced with the findings of strong interactions between WWTPs and networks for efficient management, a provisional bulletin for the management of the SIAAP networks and WWTPs is drawn up each week. It is established in liaison with WWTPs and network operators and is distributed to all operational departments and functional departments responsible for operational reviews and monitoring of the natural environment.

This bulletin provides summary information on the availability of each of the WWTPs and networks according to the work and incidents. This situation is given for the current day, with a forecast of the contributions and the distribution of flows for the next 10 days, and takes into account the annual shutdown programme for the next three months. This forecast is based on the non-real-time version of MAGES. The information is correlated with data from the natural environment (flow and temperature of the Seine) as well as trends for rain over the upcoming 10 days, and can if necessary be adapted to limit the impact on the environment. A synthetic map indicates the network shutdowns for maintenance or reduction of the capacity of the WWTPs. The situation of H_2S and CH_4 gas risks in the networks is also described.

An adapted flow management for the Seine-Aval WWTP

Of the six WWTPs, the Seine-Aval plays a major role due to its capacity and it location in the network. It is frequently used to help other WWTPs. The converse is less common because only two plants can contribute to reducing, even in limited proportions, the incoming pollutant loads on the Seine-Aval.

Low flow rates are critical for this facility, which was originally designed to handle daily volumes of 2.1 Mm^3 per day or more. In summer, flows sometimes drop below 0.8 Mm^3 per day, and rates of less than 10 m^3/s are problematic. A procedure for real-time regulation of flow rates to avoid going below this threshold has been put in place. It relies on a transfer of water from the Seine-Centre and Seine-Grésillons plants, supplemented by storage in the drainage network to support the night flow.

The main sewers that feed the Seine-Aval WWTP were originally designed to drive 2.7 Mm^3 per day. With the reduction in capacity, the low flows have made them susceptible to deposits. The recovery of these deposits during heavy rains poses operating problems that can lead to unacceptable exceedance of standards. To limit these risks, the operator requests a specific regulation with a rise in flow per stage of 5 m^3/s in time steps of 15 minutes, allowing him to change the configuration to physicochemical treatment of the stormwater in place of phosphorus.

Lessons learned

The development and implementation of MAGES are the result of a long process which started at the beginning of the 1980s and ended in 2008 with the commissioning of this system; this means more than 20 years of investment in research and engineering in the field of real-time control. During that time, a strong background has been built on this experience by SIAAP and its partners. This common technical culture is one of the key factors in its success. Even if the goals of each system were different, and even if the culture was specific to each partner, this shared technical background was very helpful to work on the integration of each system, not as a single one but to make them interoperable (i.e., the systems are interconnected and can share information and data with one another). This is the other key point: that each of the parties involved in the project kept its real-time control system on its own. The project of integrating the different systems is about connecting them together, not merging them into a single system.

In other words, when the MAGES project started, the context was mature. This does not mean that it was an easy project, especially regarding the aspect of human relationships between partners.

Four things helped to overcome these issues:

- The strong commitment of the top management of each partner;
- The technical challenges of the project, which made it exciting and motivated everyone;
- That the project started with two studies with concrete outputs for all the parties: rainfall data treatment and setup of real-time data-sharing platforms between SIAAP and each of the partners;
- And the financial support from the Water Agency and the Île-de-France Region.

The outcome for SIAAP's operators of MAGES's implementation is largely positive. In particular, it allowed:

- Developing advanced skills in hydraulics and urban hydrology in relation to the complexity of certain works and the geographical extent of the SIAAP transport system;
- Developing a new relationship with its partners, to federate the management of the Ile-de-France networks around shared objectives;
- Taking into account the overall operation of the sanitation system managed by SIAAP and the sharing by network operators and WWTPs of a common tool;
- Particular attention to the WWTPs, with a sharing of knowledge of the constraints – the setting up of the weekly operation bulletin is an illustration;
- Growing interest of the WWTPs in the forecasts by MAGES of flows arriving at the plants;
- Having a powerful tool to assist in the planning of works shutdowns and the updating of the master plan.

Of course there are possible improvements for the tool. Among them, there is one target of the project which has not been achieved: the implementation of a user club. This was one of the aims of the 'replay' mode of MAGES, and in particular using it as a tool for feedback on experiences. This would move one step towards the creation of a common culture among the partners in the operation of the sanitation system.

Next steps

Integrating the recommendations of the updated sanitation master plan

In 2017, an update of SIAAP's sanitation master plan focused on the achievement of the Water Framework Directive objectives was adopted. This plan has confirmed the need for storage facilities in order to significantly reduce the pollutant loads discharged to the Seine by the main CSOs and the Seine-Aval WWTP.

To achieve these goals three works are recommended – in order of priority:

- 100,000 m^3 of storage capacity at the La Briche CSO;
- 70,000 m^3 of storage capacity at the Clichy CSO;
- 500,000 m^3 of storage capacity at the Seine-Aval WWTP.

These works will be implemented based on a step-by-step assessment of their impacts on the receiving water.

These works and their management rules will be an important evolution for MAGES for an optimal operation of the sanitation system in order to reduce the stormwater impact on the Seine, including their impacts on WWTP performance.

From dynamic flow rate management to integrated pollutant load management and impact on the environment

Dynamic flow management is one of the ambitious and innovative projects led by SIAAP. Despite the difficulties inherent in this type of project, since 2007 the system has been fully operational, and it has become SIAAP's control tower.

As a management tool for a complex system, MAGES is far from maturity, as the constraints related to effluent pollution and the impact of discharges on the natural environment are not integrated in its current version. It offers significant development prospects, and the next steps of the work in progress aim to prepare this development.

One of the important developments for the future is the management of a system that is increasingly complex and more and more responsive, with a need for permanent performance, because of increasing pressure on the whole system related to future population growth and the projected effects of climate change on flow rates. This means a reinforcement of the constraints, reducing in some ways the 'right to make errors'. To make the most of it, more data collected by a more developed monitoring system will have to be processed in real time to take into account the changes in the pollutant load and the acceptability of the environment, while controlling the cost. Regulatory bodies require compliance with the discharge permits for WWTPs and CSOs. The data processing tools are the essential complement to help the operators make the most of the equipment implemented by SIAAP. In this sense we are getting closer to 'smart systems', or systems capable of adapting to changes in the environment.

This evolution of SIAAP's smart system relies on three complementary fields:

- Developing an evolution of MAGES that will make it able to deal with pollutant loads instead of only flow rate;
- Implementing decision support models in the operation of the WWTPs; and
- Connecting MAGES to a Seine quality model to provide a river quality forecast to adjust wastewater treatment performance to the needs of the receiving water and meet environmental quality standards.

From flow rate management to pollutant loads

Upgrading the system from flow rates to pollutant loads is still a challenge that will require an ambitious change in the system. The objective is to build a numerical model of the pollutant transformation during sewage transport in the sewer, taking into account solid transport and dilution of some pollutants during rain events. The first step will be to collect data to understand the processes that are occurring in the sewer, before proceeding to numerical modelling. This will require the development and implementation of continuous monitoring systems in sewer networks.

Sewage quality monitoring

In some situations, the system is subject to transitory phenomena at the WWTP inlet. Their origin is elsewhere, further upstream on the network. In the context of performance, the incidence of these particular situations was moderate. Today, with penalties on performance and ambitious quality goals, things are different.

The variations in pollutant load at the inlet of the treatment plants is often the origin of malfunctions. Understanding these variations is often difficult, due to lack of data. Consider for example the rapid load variations at the inlet of the Seine-Aval WWWTP during rainy weather (Figure 14). These increase the risk of uncontrolled discharges and therefore of noncompliance with regulations. These are most likely to occur due to the accumulation of solid deposits during dry weather and their flushing during rain events. However, without

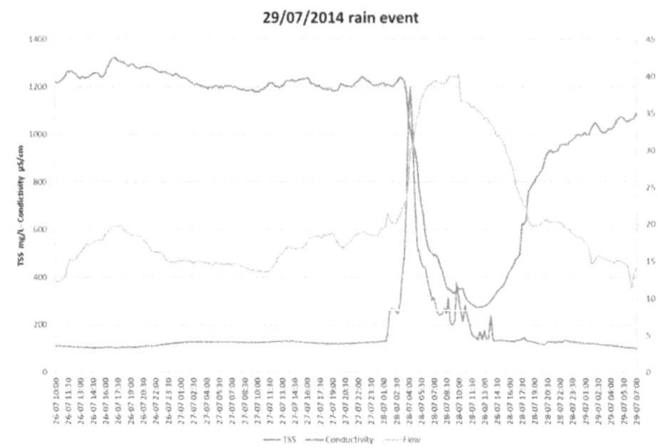

Figure 14. Real-time monitoring at Seine-Aval inlet (flow blue, turbidity green, conductivity red).

adequate instrumentation, the information required to understand the dynamics of the situation or the respective contributions from each of the five sewers to the treatment plant is unavailable. Only a continuous measurement of the turbidity from each sewer would help us understand the situation and guide any corrective measures. On the basis of this knowledge, provided by the implementation of continuous monitoring upstream in the networks, it would be possible to consider the development of a model of the deposits and their flushing, enabling a pollutant load forecast to be provided.

These metrological developments are an integral part of the MAGES evolution towards a 'MAGES pollutant flux'. The implementation and deployment of continuous metrology in the sewerage networks is a difficult but indispensable exercise to ensure that the operational knowledge of the complex functioning SIAAP network is available. This complexity is related to the extent and diversity of its structure, which includes separate and combined sewers, variable pollutant concentrations over time, and diverse industrial activities. Thus this level of detail is a prerequisite to ensure the success of a reliable pollutant load forecast. The deployment of the instrumentation will be introduced progressively to take advantage of the feedback from experience at the Clichy CSO's pilot site, which has been equipped with continuous monitoring instruments.

Treatment plant instrumentation

Another instrumentation development concerns the WWTPs. It is an important link in the system which is of increasing concern to operators. The most recent units now have continuous metrology, and are developed to run more complex plants with reactive processes. Optimized management, in particular for the Seine-Aval WWTP, requires acquiring all of the knowledge base necessary to build the expertise needed to face all of the operating difficulties. The instrumentation installed in the plant will be an indispensable tool for piloting this solution, and for knowledge acquisition. These data will also feed the development of process modelling tools and the construction of precise and efficient control loops.

Metrology in the natural environment

At the other end of the chain is the receiving water, where similar needs exist. In the framework of the PIREN-Seine, a research programme on the Seine, SIAAP's R&D

department is conducting studies to develop autonomous and continuous measurement systems for the quality of the Seine, known as CarboSeine. These autonomous systems are intended to give an accurate picture of the quality of the Seine, which is also subject to continuous variations in quality. Here too, knowledge of these variations is essential to understand the proper functioning of the Seine ecosystem and the dynamics of evolution of the behaviour of pollutants released by the SIAAP facilities. In the future, CarboSeine will also be an indispensable complement to the real-time integrated MAGES system, coupling MAGES to a Seine quality numerical model. These data will be used as self-correcting data sets for the model and also for validating the good management produced by SIAAP.

In addition, the feedback of these data to MAGES and to the WWTP operators will give a real-time picture of the quality of the Seine and of its possible fragility, highlighting the daily context of their performance objectives. These data will also make it possible to report to the state authorities, or even potentially to users, the real impact of the sanitation system on the environment.

Developments in modelling

There are three areas in which modelling will continue: the transport of pollutants in the SIAAP networks; the operation of biofiltration treatment plants; and the quality of the Seine. In each case the models to be developed will largely rely on reliable and good-quality metrology. As mentioned, the situations that must be managed are more and more complex and variable, with variations in the distribution of the flows between plants, management of increasingly complex rain weather, and the future presence of stormwater basins which will require volume management. The quality of the Seine is variable too, but to a lesser extent.

Flow modelling

The modelling of the transport of pollutants in the SIAAP networks is an innovative project. Initially the modelling will concern the deposits of suspended matter in the main sewers of the Seine-Aval WWTP. In addition, due to the continuous measurement of the sewers, the behaviour of pollutants in certain sectors of the SIAAP network will be measured, so even simple modelling such as the dilution of ammonium in rainy weather, may be considered.

In terms of treatment process, biofilters are very reactive, with response times of an hour or so. This is an asset, because they adapt well enough to the rapid variations of flows arriving at the Seine-Aval WWTP. However, it is also a constraint, as any operational error can result in significant degradation of the discharged water quality. Therefore, SIAAP is operating an increasingly complex sanitation system while being more responsive, with narrowed margins of error. This situation is particularly true at the Seine-Aval WWTP, whose regulatory functions (or buffer of the western part of SIAAP's system) are vital for the proper functioning of the entire sanitation system.

To address this, SIAAP's R&D department is partnering with the University of Laval in Quebec and IRSTEA as part of the Mocopée research programme (www.mocopee.com) to model the operation of biofilters. The aim of this programme is to build models for predicting the operation of integrated processes in the sewage treatment in our plants. Unlike activated sludge treatment, which has been the subject of scientific studies for over

20 years and for which there are good models, there are few modelling tools available for physicochemical settling and biofiltration processes.

These models are constructed to predict the impact of changes in operating conditions (reagent injection, air injection, applied flow rate, applied pollutants loads, etc.) on the quality and performance of the treatment. The use of such tools aids in controlling the processes – in controlling residual concentrations at the outlet, but also in controlling energy and chemical reagent costs. The models should be seen as an aid for piloting the plant, not as automated tools. Coupled with a pollutant load forecast and a performance directive, these piloting aids will assist the operations management with their installation facilities.

Taking into account the quality of the Seine in MAGES in real time

Another major development of MAGES is its coupling with a Seine quality model. This option was planned from the very beginning of the MAGES project, and implementing this step is critical as it is the missing link towards integrated management of a sanitation system. This would be a very important step forward in sanitation management. This project will make it possible to define the downstream sanitation system management constraint in an intelligent and adaptive manner according to the environmental requirements.

Thanks to the PIREN Seine research programme, a detailed deterministic model of the Seine quality has been produced, Prose, which will provide the base of a dedicated model adapted to real-time control.

Conclusion

Twenty-five years after the implementation of SIAAP's first real-time control system, SCORE, and 10 years after commissioning a major upgrade of this real-time control system, one can see how far it has come, and several lessons can be learned.

Through this example it appears that sanitation is seeing the same evolution as many industrial sectors in the search for security and efficiency: automatization and numerical modelling are more present, as sanitation transitions to a mature industry. This appears to be relevant and necessary as risks are growing larger than ever, especially from an environmental point of view: the improvements of the receiving water quality and the public awareness of environmental issues are making failures in wastewater treatment unacceptable. In these conditions, a failure can be close to an industrial accident. We also have to take into account the fact that regulations are becoming stricter than ever before, and in the end the 'right to fail' is no longer acceptable. Another aspect related to the transition to a real industrial approach to sanitation is the need for efficiency by searching for the best compromise between cost and performance. Real-time control can reduce the need for new investments thanks to optimized operation of the facilities.

The evolution of the real-time control system is the result of a process in which the input of long experience was one of the major elements for a successful story. This is particularly true if one considers the ambition of the project:

- This is a large and complex sanitation system, combining the different approaches of five operators, each department of SIAAP, and SIAAP itself, in a sewage transport system which mixes combined and separate sewers.

- The large extent of the territory is also a key issue, especially given the diversity of rain conditions from place to place.
- Five operators means five cultures and five technical systems that have to work together, but one of the key factors for success was that the project was driven with the idea that each operator would keep its system and not merge them together in one system.

Ten years after MAGES' commissioning, three lessons can be drawn from it:

- Thanks to the information sharing between all the operators involved in the management of the sanitation system, everyone knows that they are working for a unique system. The Paris flood events of 2016 and 2018 showed that it is not only a mindset but a reality.
- In a sanitation system which relies on six interconnected sewage treatment plants, MAGES played a major role in the necessary shift to a global overview of the operation of the system.
- The organization has evolved to adapt to these new conditions. This has been particularly true for the maintenance works programme. Now a three-year coordination programme for maintenance work is developed and updated in real-time. This system allows that when shutdown of works are necessary, exceptional discharges of raw water can occur. A weekly bulletin gives forecasts for the coming week on operational conditions: the available treatment capacity of each plant, ongoing works, but also the sensitivity of the receiving water to pollution or a general meteorological trend.

The future of the system is the next point of focus for project managers. The new developments are focused on several fields:

- Strengthening the relation between sanitation system operation and its impacts on the receiving water in order to follow the Water Framework Directive with better operational costs;
- Preparing the transition from a system based on flow management to a system which also takes into account pollutant loads;
- Introducing operation costs as management criteria; and
- Preparing for new demands, such as swimming in the Marne and Seine Rivers.

These evolutions are required to prepare for a future which will be more constrained due to the growing population and the effect of climate change on the Seine flows, which are projected to be lower. As in the past, when engineers and decision makers invested in innovation, one has to prepare for the future with smarter tools for a smarter sanitation system management to fulfil expectations of a 'water-wise' city.

Notes

1. Oceanic regimes are a characteristic of Western European climate: under the influence of the Gulf Stream in the Atlantic Ocean, the climate is cool, with high rainfall in winter. Low evaporation added to rainfall means more runoff and a higher flood hazard.

2. *Seine Grands Lacs* is administered by the departments of Hauts-de-Seine, Seine-Saint-Denis, Val-de-Marne and Paris, which own the large lakes upstream of the Paris region.
3. Readers of the printed article can view the figures in colour online at https://doi.org/10.1080/02508060.2020.1830584
4. Hypothesis adopted by SIAAP for the 2015 revision of its sanitation plan, in agreement with government authorities.
5. WWTPs are based on traditional activated sludge with high-level performance, in particular in relation to the major parameters such as biochemical oxygen demand and ammonia. Concerning nitrate, the current regulation obliges domestic WWTPs to remove 70% of the total nitrogen. Today the main nitrate contributor is agriculture.
6. The 'western system' is the downstream part of SIAAP's sanitation system. It is based on the Seine-Aval, Seine-Centre and Seine-Grésillons WWTPs, which are interconnected and on the west side of the Paris region.
7. In French, *dossier de consultation des entreprises.*
8. Deterministic modelling in this case refers to all the physical process involved in the flow calculation at each point of the network, relying on mathematical formulae to describe the hydraulic conditions, such as slope, shape of sewer, sewer material with Strickler coefficient, hydraulic formula for each kind of device, and so on.

Disclosure statement

No potential conflict of interest was reported by the authors.

References

Blanchet, B., Fradin, A., & Tarif, P. (2008). Outil d'aide à la gestion dynamique et coordonnée du système d'assainissement de la région parisienne. *MAGES (Modèle d'Aide à la Gestion des Effluents du SIAAP) – TSM*, *12*, 55–67.

Tabuchi, J.-P., & Blanchet, B. (2016). *Les apports de la gestion automatisée à la gestion du système d'assainissement de l'agglomération parisienne – In Territoires en transition, Mettre l'intelligence numérique au cœur des services, ouvrage introductif - 95ème congrès de l'ASTEE, [128 – 133].* p 178 https://fr.wikipedia.org/wiki/%C3%8Ele-de-France

Tabuchi, J.-P., Tassin, B., & Blatrix, C. A. (2016). *Greater Paris water and global change, water megacities and global change, Portraits of 15 emblematic Cities of the World.* UNESCO/ARCEAU, p 40. http://www.arceau-idf.fr/sites/default/files/FR%20-%20Paris%20monographie.pdf

Part III

Innovative Uses and Critical Perspectives

Is Smart Water Management really smart? What experts tell us

James E. Nickum, Henning Bjornlund, Raya Marina Stephan and Stephanie Kuisma

Following the precedent set in the first IWRA special issue, on wicked problems facing water quality management (Nickum et al., 2018), we posted the following Question(s) of the Year 2019 on the IWRA website: 'Is Smart Water Management really a smart idea? If so, in what ways? If not, why not? Or is it smart in some ways and not in others? What could make it smarter (institutions, policy, other)?' Seventeen people responded with very thoughtful responses. We offer a selection of these here to show the range of views. Some have been edited for clarity, and/or condensed. You may find the originals (including those omitted here) at www.iwra.org/questionoftheyear2019. We have added headings which we hope accurately reflect and contextualize the opinions of the respondents.

Like the present special issue, this solicitation of answers from members is part of the IWRA's Smart Water Management Project. For more information, visit www.iwra.org/swm-2.

Smart Water Management (SWM) will not answer core questions about priorities and values

SWM means different things in different contexts since it is not a precisely defined term. If it is broadly defined to mean the application of technologies to manage water resources then, like any tool, SWM works best where a long-term vision and purpose have been laid out in advance of investments in the tools themselves. With changes now occurring in the water cycle due to climate change, water management challenges extend beyond the traditional objectives of improving efficiencies to rethinking water use and infrastructure design, as well as recognizing the co-dependence, with humans, of ecosystems on sufficient water supplies to sustain themselves. These challenges require ongoing adaptive approaches and SWM can assist with numerous implementation and monitoring functions but it will not answer the core questions about priorities and values.

M.P. Trudeau

SWM has a bright future in urban areas in developed countries; some work needs to be done to extend it beyond that

Population growth, modernization, and all other areas of development require that available resources be used efficiently. Being able to control and track water use with technology is a good idea, though it cannot be applied in most rural communities. It is obvious that in developed countries a lot of successes will be recorded with SWM, unlike most developing countries. This concept, in the future, must meet global needs, and this can happen when it is understandable, cost-effective, easy to use, and sustainable.

Peter Addo Amarkai

Water systems have many social objectives; can SWM help?

SWM is a great concept, but it has many facets and must be explained that way. For example, it can make systems more efficient, but like other smart technologies, it has concerning implications for privacy. Also, water systems have many social objectives, and the jury is out about whether smart technologies help with those. It is my opinion that smart technologies can help with them, including helping with access and affordability of drinking water.

Neil Grigg

The need is there, even in some of the least developed, most arid countries: ask Chad

SWM is a solution to address water losses in a distribution network. According to the Chadian Water Company, 40% of the total volume of water produced per day is lost to leakage, pipe bursts and illegal connections to the network. This constitutes a great economic loss of the country and decreases the efficiency of water supply to the customers. Under these circumstances, information and communication technology (ICT) can help improve water service to customers. If 40% of the water loss is expressed in terms of cash over a period of 40 years, it would be enough to adopt an ICT system to monitor the distribution network. For sustainable water management, SWM is needed in all African cities with pipe-borne systems as they face climate change, population growth and continuously increasing water demand. SWM is feasible, with the support of institutions and strict policy implementation.

Ramadan Marne

Transboundary waters present challenges for SWM

SWM is a very good idea in the context of transboundary water management. Yet it confronts many challenges on the regulatory and institutional levels as new information and communication technologies result in the emergence of new data that may change the positions of the different countries or affect the amount of water they may receive. Existing regulatory frameworks for transboundary water agreements may pave the way for the establishment of joint institutions, and have provisions for data and information exchange, but they lack elements that take into consideration new technologies and changes that may occur. This reality will affect the existing agreements and institutions as well as the future ones. A similar situation may be already witnessed in the context of big data, which I have been researching in the past year. Places like South Africa have established specific projects to deal with the impact of new technologies on the management of transboundary water resources, be it surface water or groundwater, but gaps remain in the legal frameworks.

Imad Antoine Ibrahim

For SWM to be successful, it has to be more than fancy gadgetry

Some drivers of water insecurity, such as rapid urbanization and climate change, are hard to control. Water utilities are more manageable. SWM can play a critical role, but the key is to transform these operators into modern and efficient organizations, valuing the people behind them while putting smart systems in place.

In recent years, the development of more affordable sensors, connected devices and communication channels has revolutionized the quantity of technical data, essentially collecting from the ground information on water flows, pressure and quality. Integrating this data from different sources on a platform has improved the management of infrastructure, energy and manpower.

According to a 2016 Global Water Intelligence report, the market for control and monitoring solutions in the water sector worldwide is expected to reach \$30.1 billion in 2021, with Asia and the Pacific having the biggest market share, at \$10.3 billion. No wonder automation or the greater Fourth Industrial Revolution is gaining traction – smart systems will only progress further and increase quality of life. Of course, smart systems are not magic. The goal is not to have or use fancy gadgetry but to optimize the operations of the water utilities and raise their service delivery so that the water network and infrastructure work for the people. …

Three focus areas – infrastructure, operations and institutions – together, made more formidable by smart systems, build a solid base from which we can help ensure water supply and equitable access. This is just the start, but it could be coupled with water conservation measures, integrated resources management, and proactive governance.

This gives us a chance to avoid potential water crisis and conflict. At the end of the day, what we want is water for all, 24/7.

Stephane Bessadi

SWM, like IWRM, is a good idea but the key is correct implementation

SWM is of course a good idea – it should improve efficiency and decrease costs – but as happens for all good ideas (see IWRM) the issue is correct implementation.

Cases studies show that the application of new technologies may lead to a substantial improvement of water resource management as well as improved access to water and sanitation. But these technologies need a capacity to be correctly used: this capacity relies, on one hand, on the ability efficiently to manage databases (collection, validation, treatment, easy access for the users) while, on the other hand, ensuring people have the capacity to make use of the possibilities offered. SWM must include a strong component of training and capacity building, as well as appropriate communications involving the people concerned to ensure social acceptability.

Guy Fradin

SWM has to be smarter than IWRM was

SWM as a concept evolves in response to the rising need for collaborative information sharing and practices in water resources management. Automated responses towards increasing security and efficiency, while decreasing risk and uncertainty, may be the add-on that SWM brings to Integrated Water Resources Management (IWRM) and the discussion on water resources management. IWRM typically relies on principles of common pool resource management with relatively clear data-driven demand forecasts, where all players, as user sectors, are given voice and apportioned user rights. But as a concept, IWRM is often criticized for both excess and lack of adherence to scientific principles, questioning its value outside academic discourse and often described as a consensus without practice.

The assertion that real-time data and automation will automatically translate to more efficient services, more reliable water management, inclusive decision making, and improved collaborative knowledge sharing still banks on the idealism that befell IWRM as a concept. How can SWM facilitate access to data, especially in transboundary basins? How will AI-driven automated systems bypass the need for self-sufficiency in water security to ensure equitable distribution with zero conflicting political interests? In addition to IWRM being regarded as ambitious in its objectives, idealistic in its agenda, and only used to secure funding, it is thought to have by now achieved measurable impacts, the absence of which degrades it to a 'hollow concept without merit'. It is therefore easily envisaged that as SWM inclines more towards socio-ecological consideration and institutional reforms in the planning and implementation process, it will open avenues for similar criticisms that waylaid IWRM.

The mention of water in IWRM drives a perception that it is single-faceted and non-holistic, as it will be illogical to integrate a single resource. Despite an origin focused on a water–society interface, IWRM has struggled to evolve into a balanced account of human–environment relations and to include consideration of the entire ecosystem. This is what SWM as a concept must watch for as it carefully creates a niche within the confines of IWRM by first capitalizing on its deficiencies to provide proven, tangible and trackable progress before being embedded within. Summarily, SWM can be smart in its own way by first evolving as an independent provable concept that capitalizes on real-time data and the need for automation in the water sector before being embedded within other existing concepts.

Amali Abraham

ORCID

Henning Bjornlund http://orcid.org/0000-0003-3341-5635

References

Nickum, J. E., Bjornlund, H. & Stephan, R. M. (2018). Wicked problems facing integrated water quality management: What IWRA experts tell us. *Water International*, *43*(3), 336–348. doi:10.1080/02508060.2018.1452879

Smart water management: can it improve accessibility and affordability of water for everyone?

Neil S. Grigg

ABSTRACT
By improving the operational capacity of utilities, smart water systems may enable utilities to improve access to water services and their affordability. Use of smart technologies is increasing in leading-edge utilities for this purpose. However, water access may be problematic where utilities perform poorly or when utilities do not choose to extend universal service, and smart tools can address both problems with increased effectiveness of operations and improved customer interfaces. In either setting, effective governance is required, and security and privacy issues will be of concern. To explore the feasibility of smart tools, demonstration projects will be needed with political and financial support. An example of using smart tools to facilitate service extension in collective housing is presented. While this is one way to use smart tools, other pathways are available, especially their use to improve overall effectiveness of utility operations.

Introduction

Smart water management (SWM) systems offer new choices for the ways that water purveyors operate and consumers access their water services (Christophe, 2018; Karwot et al., 2016). These choices provide new capabilities to address persistent problems of access to water services and their affordability. SWM systems are a subset of emerging applications in smart cities (Eremia et al., 2017; National League of Cities, 2016; Pereira et al., 2018), which promise to create more functional and livable communities through use of information and control technologies (ICT). Smart cities feature innovations in services such as use of social media for community information and critical functions such as emergency alerts.

Within smart cities, SWM systems mainly address the operations of the water utilities, which provide unique environmental and health services. They can also operate at different scales and in different management scenarios and stretch to embrace the full integrated urban water system, which can include features such as water recycling and stormwater. This article examines how emerging technologies will affect two central issues of the water industry: access to water services for all sectors of society; and making those services more affordable in the face of rising costs and pressures. The potential to use SWM to address these vexing problems is intriguing because billions of people globally lack access to safe, reliable and affordable water services; and progress, while encouraging, can be slow and not help everyone on an equal basis. Meanwhile, lack of access to affordable services hinders the convenience of daily life and threatens the public health of billions of people globally.

The discussion begins with the underlying problems of water access and affordability as they are affected by the capacities and technical problems of water purveyors and by their business decisions about access for low-income customers. How the technologies of SWM may affect operations, change customer interfaces, and mediate problems of affordability is discussed next, and examples of how they can affect the capacity of water purveyors to address access and affordability are analyzed in three common problem scenarios of utility capacity and culture. The emphasis in one example is on how smart technologies could enable extension of water services in collective housing. The conclusions offer a path forward to exploit the potential of SWM to address the problems of access and affordability in the three scenarios.

Problems of access and affordability

Metrics of the many people lacking access to safe and affordable water supplies are available, but the numbers are uncertain (Grigg, 2018). The Joint Monitoring Programme (2017), a joint effort between the World Health Organization and UNICEF to track access to water and sanitation, indicates that 4.2 billion people have piped water on premises, which is an indicator of access. However, the actual number may be much lower because definitions are hard to nail down and data collection is difficult. As an example, Truelove (2018) showed with field research that official reports in India are biased towards overstatement in that they indicate high levels of access to water, but the reporting system is unreliable due to uncoordinated measurements and bureaucratic practices that ignore actual problems. Regardless of the numbers without access, to mitigate the situation requires good understanding of why access is unavailable to people and of the possibilities to overcome the barriers.

Access to water services is denied to many people for various reasons, chiefly utilities' capacity and customers' ability to pay. Ideally, new connections could be added to community water networks to accommodate growth and sustain near-universal service levels, but this does not work in settings where people live in unconventional arrangements, lack property rights, have low incomes, and may live in slums. Reasons for lack of service connections range from dysfunction of water providers and customers' inability to pay, to unwillingness of local authorities to extend services (Hunter et al., 2010). Water provider dysfunction can be addressed by performance improvement. Lack of ability to pay is an issue of poverty and may in some cases be mitigated by developing low-cost basic services and rate adjustments that take into account local situations. Utility unwillingness to connect people may be due to political issues, such as not encouraging permanent settlement, which can be addressed by offering temporary services that do not involve high capital expenses when low rates of consumption are ensured. These problems occur in all countries, but in different ways, and problems must be studied in their contextual settings to understand the problems of access. In general, access will be more widespread in high-income countries than in low-income countries, and affordability issues will be driven by the level of poverty in each country, while being more widespread in low-income countries.

The Joint Monitoring Programme estimates are based primarily on household surveys, where water users occupy varied degrees of housing, defined as conventional dwellings, basic dwellings, temporary housing, marginal housing, and collective living quarters. In all of these, four water-dependent support facilities are deemed essential: drinking water, cooking, bathing and toilet. By definition, conventional housing has all four, basic and temporary housing may or may not have them, marginal housing does not have them, and collective housing usually has shared facilities (UN Department of Economic and Social Affairs, 2001).

In all countries, water access is linked to the quality of housing arrangements, which depends on household income and indicates levels of access and affordability. The other main variable is the effectiveness of utility operations to address issues such as equitable rate-setting and control of water losses. Taken together, effective utilities and conventional housing generally represent scenarios of people living in cities with established services and relatively minor financial problems, and the less-capable utilities serving people in marginal or collective housing are generally at the other end of the range, with lower access and greater affordability problems.

Low-income countries generally have larger percentages of populations in basic or marginal housing, and a billion people still live in slum conditions, with the absolute numbers continuing to grow (UN Habitat, 2016). It follows that many people in these countries live in substandard housing on the lower end of the scale and often lack access to numerous basic needs, including adequate water supply.

The pathways to provide piped water to different types of housing are by providers or by individually supplied systems. Providers offer connections through single-family, multi-family and consecutive systems (where a utility or user provides a connection to another utility or user). Self-supply is from wells, standpipes, tank trucks, bottled water, rainfall harvesting or illegal connections.

Many water providers are public utilities, which vary from huge, well-organized and high-performing ones to basic small systems. A typical country will have a few large utilities and many small ones. These may be classified by the capacity of the utility to address a range of technical, managerial and social issues (Hopkins, 2018).

To provide a view of the pathways to access to water, Figure 1 shows utility status and customer status according to housing. The terms 'nascent', 'emerging' and 'developed' were used by Cohen (2016) to refer to utilities just starting, those still needing development, and those that function at a higher level. 'Consecutive service' means that residents can pass along their water service to other water uses in collective living. It is adopted from the common situation where one utility gets water from another utility rather than from its own sources. Another form of consecutive service is a condominium, where one water tap serves a number of units within a complex (Melo, 2009).

Using the variables of housing quality and utility effectiveness, a matrix can display contextual situations of access (Figure 2). The upper-right quadrant is the high-income-country case, where access is mostly available and utilities are mostly effective. In the lower-right quadrant, utility services are effective, but people in low-quality housing may be denied access. In the upper-left quadrant, conventional housing in small settlements may not have access for lack of utility capacity. In the lower-left quadrant, utility services

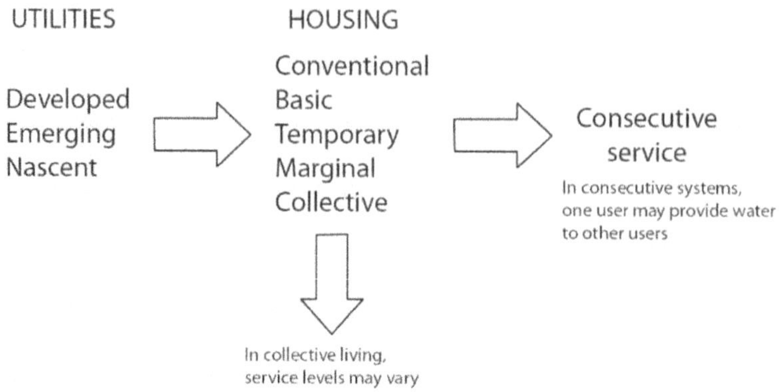

Figure 1. Utility effectiveness and housing arrangements related to water access.

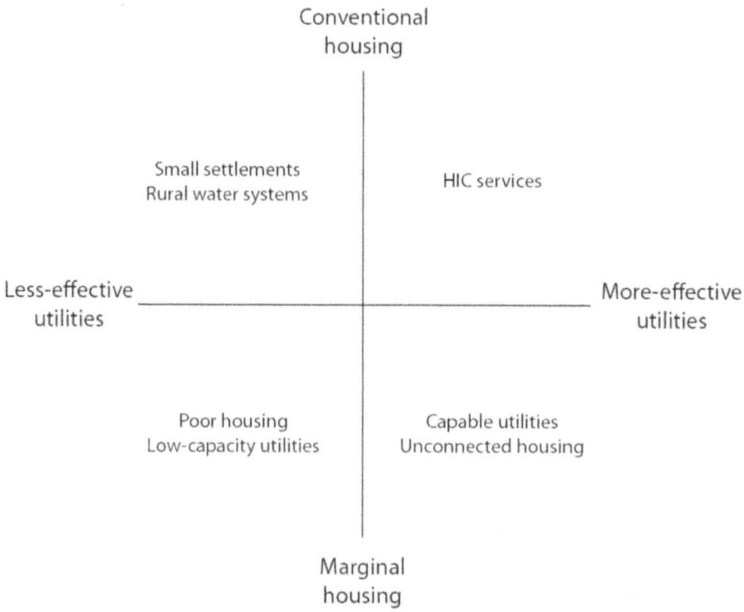

Figure 2. Cases of water access by housing and utility status.

and housing quality are both inadequate, which creates the most difficult example of access and affordability. The three cases of interest for further analysis are lack of utility capacity in small settlements, lack of utility capacity in larger cities, and unconnected housing where services are available for others.

In small communities, access may be limited because utility services underperform or cannot be extended to everyone. This is the reason, for example, that in the US rural water systems are sometimes subsidized. How SWM could help will depend on whether it can compensate for low utility capacity and resolve the financial and other problems of extending services to remote areas. With low-capacity utilities in larger cities, service problems occur regardless of housing level. Low-quality housing is likely to receive less attention than higher-quality housing. The major issue there is utility capacity, but different from smaller communities because the scale of the problem is larger and people may cooperate less in larger cities than they do in small communities. If the utility has high capacity but does not connect all housing, the problem may involve the business model rather than management capacity. This is a common problem when utilities are managed as enterprises and must break even financially due to lack of subsidies.

SWM tools to address system operations and access to services

SWM tools are emerging for many applications. For example, a case study report from the International Water Resources Association (2018) has 10 cases, ranging across several purposes of water management. Those SWM cases that deal with urban water supply targeted real-time monitoring of drinking water quality, leak detection, energy optimization and customer satisfaction.

The main way that SWM tools can affect access to water is through interventions to increase the effectiveness of operations and to improve the customer interface with water services. These two aspects of SWM are shown in Figure 3 as 'operational results' and 'customer access'.

The diagram shows the water system on the left, with ICT tools employed to provide the smart technologies for its management. System performance metrics are displayed on an operations dashboard, which is common in developed utilities. Also becoming more common are customer dashboards, which can be accessed by water users as they interact with the water system through websites. Different types of customers can access the dashboard, including those who receive services consecutively from a primary user. Examples would be residents of an apartment or people living in condominium housing. In the future, these distributed cases could be addressed by the 'Internet of things', providing the capability for real-time monitoring and control of individual fixtures to facilitate time-of-day pricing, for example, as in electric power utilities.

Operational interventions for SWM are developing in parallel with those in other infrastructure control systems, such as smart electric grid, real-time traffic control, flexible public transit and sustainable use of energy. The interventions with customer interfaces include features such as weather and community information and bike-sharing programmes, (Drell, 2012).

While SWM introduces new tools, ICT for water systems have been evolving since the advent of computers and even before with innovative telecommunications and mechanical controls. The field of control engineering tracks progress with tools that facilitate automatic control of mechanical systems and related ICT equipment to collect data,

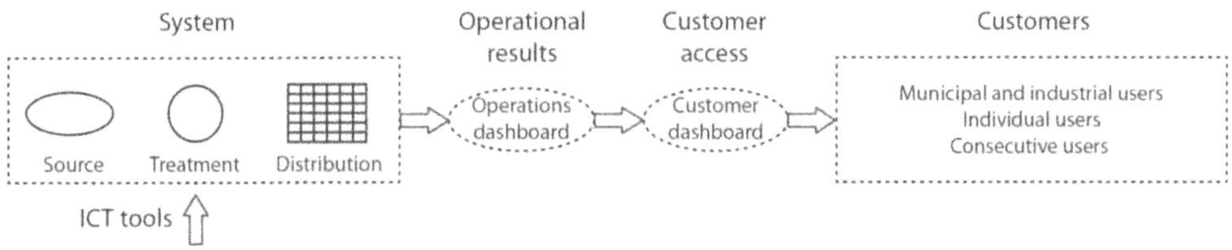

Figure 3. Operational results of utilities and customer access to water.

perform analyses, make decisions, send commands and actuate controls of systems. SWM is thus the child of legacy water control systems, especially supervisory control and data acquisition systems.

ICT tools can be viewed by their elements (data, models, mapping, controls, devices, monitors, sensors and communications) as they are used within management systems with smart command-and-control systems frameworks, such as advanced metering infrastructure, supervisory control and data acquisition systems, database and modelling software, geographic information systems, decision support systems, programmable logic controllers, and dashboards to provide access to sophisticated and integrated ICT systems.

Using these new tools, SWM technologies can improve efficiency and make infrastructure more secure. Examples of positive outcomes are to manage pressure and flow, clean and repair pipes, assess conditions, map facilities and users, optimize energy use, track water age, sense low fire flow, detect events and locate illegal uses (Norton et al., 2019).

If SWM were only about automatic control, it would not matter much to water users, other than how it affects the performance and cost of their water services. However, like other smart systems, the human interface will change significantly. SWM can create new customer interfaces, such as online water bills and real-time information on water system conditions. With SWM, citizens can participate in decision making through online engagement and access to more information, to increase transparency and trust in water systems. For the customer interface, automatic meter reading has already made large inroads in changing the relationship between utilities and customers. Data on consumption can be collected automatically and transferred to databases for analysis and billing. With automatic meter reading, no one enters private property to read a meter, so customers are disturbed less.

The effects of SWM can be felt down to the site level, where the 'Internet of things' offers new possibilities to monitor use and quality at each plumbing fixture and bill for water use at micro-levels. This is similar to energy pricing, where electric power consumption can be monitored down to the individual device. As long as water costs are low, this may be unrealistic for plumbing fixtures, but as costs rise it can offer ways to control high-water-using fixtures and water leaks.

As shown conceptually in Figure 4, ICT can be used in operations and customer interfaces to help overcome barriers such as lack of infrastructure, water loss and

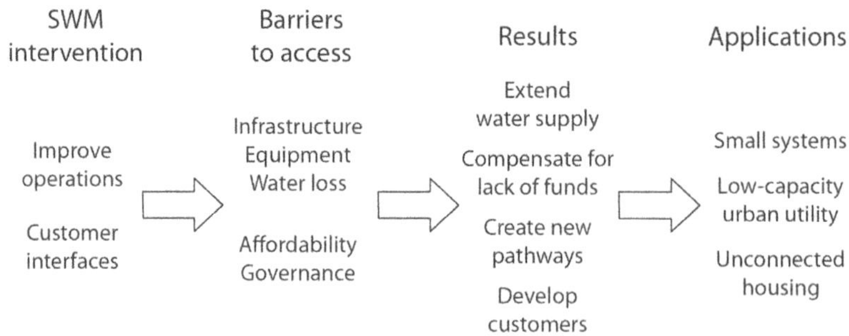

Figure 4. Smart water management interventions to overcome barriers to water access.

interruption of source of supply, and to detect problems such as inadequate water treatment. They can interact with the customer interface in identifying water users, detecting illegal connections through advanced metering, and establishing paying accounts.

Anticipated positive operational results are extension of supply by reducing losses, compensating for lack of funding to construct water infrastructure, creating new pathways to serve customers, and developing new customers such as by converting illegal connections to paying customers. Finally, the diagram points to how these will play out in the three application settings.

How SWM can affect access will differ among the three settings introduced earlier, but the general improvements apply to all settings. It might help utilities gain control of systems, reduce costs and extend services. This can occur with technologies that enable them to overcome lack of personnel, equipment or know-how to improve operations. If a utility lacks the water to serve all residents in a city, SWM might extend the supply by finding leaks and reducing losses. While some leakage is inevitable, in many utilities it is excessive, sometimes reaching half of all the water produced, or even more. If financial recovery is a major issue, SWM might locate illegal taps and either eliminate them, if the saved water could be sold, or 'legitimize' water users and bill them. There are no metrics on the extent of water theft, but anecdotal reports have it as widespread in many cities. With SWM lowering the costs of monitoring and delivery, a utility has new opportunities to identify previously unserved low-income parties and enrol them as paying customers, even at basic levels.

Affordability of water services

As technology advances, many people benefit, but the affordability issue must be confronted to meet the needs of low-income people. Affordability issues are different in high-income and low-income countries.

In the high-income countries, the cost drivers of water services are ageing infrastructure, tighter regulations and the need to make systems more resilient (American Water Works Association, 2016). Rate increases are needed to sustain high-quality service (Black & Veatch Management Consulting LLC, 2020), but their impacts on low-income customers are of major concern. This reinforces the need to reduce theft and leakage to increase the percentage of water production that provides revenue and thus reduce the need for rate increases.

In the US, for example, debates about affordability focus on rates as a share of household income. Current thinking is that 2.5% of household income is an appropriate ceiling for water supply rates (not including wastewater). A family with an income of USD50,000 per year would pay up to USD1250 per year under this guideline. That is beyond the reach of many households in the US, and system costs in cities elsewhere around the world are similar.

In the US, affordability of water services is a lively policy issue (National Academy of Public Administration, 2017). Issues focus on questions such as community versus individual affordability. Community affordability has to do with the level of infrastructure that can be financed, while individual affordability is about whether the lowest income groups are able to pay. For low-income groups, many utilities provide customer assistance programmes, but these can be challenging because many customers live in single- or multi-family rental buildings or public housing and pay for water through rent or a home maintenance fees, putting them out of the reach of utility assistance programmes. They are likely to be in poverty, have a disability, speak English as a second language, and/or have lower-than-average education levels. Even if they pay a lower total than others for water services, they pay a higher percentage of their income (Cuppett et al., 2016).

Water utilities face the dilemma that there is consensus that they should operate as businesses, but also that they should address affordability for low-income customers (Hopkins, 2018). The question is how to balance the two roles. The American Water Works Association (2016) issued a policy on affordability which seeks to address both:

Utilities should work closely with their local, state, provincial, and national governments to ensure that applicable laws and policies do not impede utility efforts to address affordability challenges and evaluate new policies that allow low-income households to have access to utility services, while maintaining the fiscal sustainability of utilities.

The association cites a flagship programme to include bill discounts for low-income families; crisis assistance vouchers; assistance to eligible customers in multi-family dwellings who are at risk of eviction; emergency plans for employment, medical and other emergencies; and assistance to repair toilets, faucets, plumbing and underground leaks.

Affordability issues occur in small and large communities, whether the utility has high capacity or not. An effective utility has more opportunity to address the issues. The questions are, can utilities use SWM to extend services to more people with lower incomes, and how would they do this? Operational improvements can help by reducing leaks and illegal connections to free up water for different types of housing. New customers can be connected by offering basic services and flexible payment schemes. With SWM to improve operations and deal with customers better, there is a real chance to improve affordability as well as access.

At the global level, the picture focuses on economic levels and incomes. Services requiring infrastructure at the same level as high-income countries are out of reach for billions of people.

Median household income globally is about USD10,000, and costs at the level of those in the US would be unaffordable (Phelps & Crabtree, 2013). There is also wide income variation within countries.

The global discussion about water access focuses on cost projections to meet the targets of the Millennium Development Goals or the Sustainable Development Goals, which are modest compared to the costs of higher-level infrastructure services. Hutton and Bartram (2008) estimated that USD42 billion is needed globally to meet the 2015 MDG water targets and USD322 billion to sustain the existing services. These costs total USD36 billion per year for a 10-year time horizon, which would extend coverage in rural areas and maintain coverage in urban areas. While investments at these levels would improve access, the use of SWM may improve services beyond the levels envisioned.

The core conflict about affordability focuses on human rights. The US Conference of Mayors (2014) called the Environmental Protection Agency's water affordability criteria evidence of 'class-based environmental injustice' because they are based on average incomes, and lower-income people struggle much more than the average to pay their bills. Issues of affordability in the US led to an investigative visit by the UN Special

Rapporteur for Water and Sanitation, Catarina de Albuquerque (2011), but such inquires do not draw much attention in the vast reaches of the US. Organization of a national campaign for affordable water in 2014 followed protests about water shutoffs in Detroit (Walton, 2016), and periodic protests of this kind continue as part of the larger focus on creating a more equitable society. Larger-scale disruptions happen in other countries, such as when protests over water rates and privatization in Bolivia overturned the government during the Cochabamba Water War in 2000.

A scenario to demonstrate SWM potential

Many scenarios of the application of SWM can be conceptualized, and parts of them are evident in cases such as those published by the International Water Resources Association. One scenario is offered in this section as an example of SWM used to improve access and affordability in a setting where an urban water utility uses ICT tools to overcome capacity problems, extend services to low-income customers, and make those services more affordable. The approach is conceptual, but it builds on recent research and uses of ICT tools in the broader water industry.

The scenario is based on the water distribution system shown in Figure 5, where the goal is to extend service to people in condominium housing in a low-income setting where service has been unavailable. The term 'condominium housing' refers here to the situation in developing countries, although condominiums are also found in middle- and upper-income settings. The same approach for extension of water services could apply to collective, apartment or slum housing. Before the extension of service, people living in the condominium housing are without piped water and must use self-supply, packaged water (bottled water or sachet water) or delivered water (cart or tanker).

The technical approach begins with the pathway for piped water from the water distribution system, an infrastructure network of buried pipes with hydrants, valves,

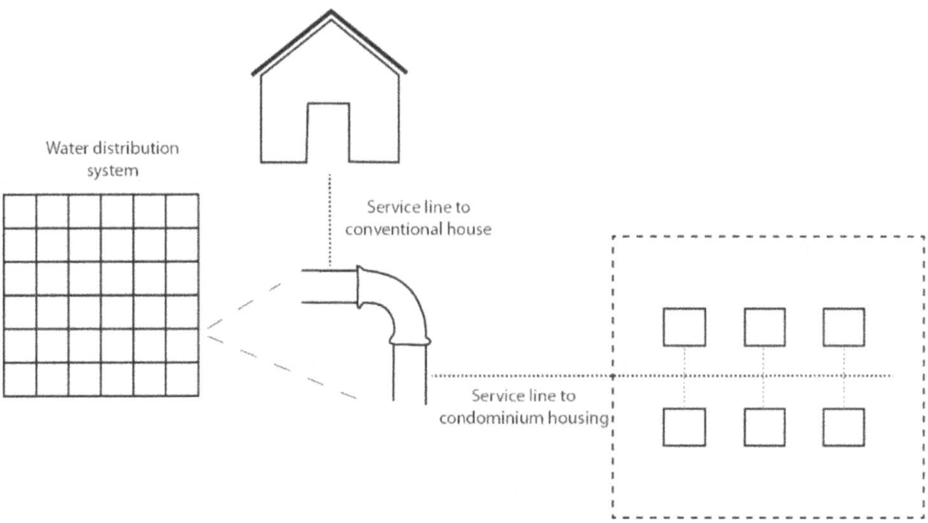

Figure 5. Water services extended to conventional and condominium housing.

tanks and other appurtenances. The usual situation is a single service line tap for a building, whether it is conventional housing, a commercial place, an apartment building or some other type.

The concept of extension of service just described aligns with water treatment innovations beginning with point-of-entry (POE) systems added at the main connection with the distribution system, and point-of-use (POU) treatment systems, which can be added at plumbing fixtures inside buildings. These were labelled 'non-conventional' approaches to water service provision in a study by the US Water Research Foundation (Raucher et al., 2004), and they have been studied as especially applicable to small water systems (Cadmus Group, 2006).

In POE systems, water is delivered to a fixed point that serves as a feeder into a home, business, school or facility. POU devices treat water for direct consumption in drinking and cooking. A POE system might or might not include a treatment device, and can be coupled with POU devices for treatment. POU devices are sold at general merchandise retail outlets and include various configurations, including faucet-mounted filter systems.

In the scenario shown, bulk water of good quality is available for distribution at POE entry points and to collective and/or condominium arrangements. The feeder pipes can be small in diameter, low in cost, and not buried if climate allows. Fire protection can continue to be available from the bulk water system. The feeder services can be connected and disconnected quickly and require little capital investment, as they are temporary water services installed by utilities. If the concept is to provide water to temporary housing, the lines can be low-cost plastic and run on convenient routes, with little construction required. Users can decide whether to use POU systems, which can be as basic as simple filters or more advanced to include reverse osmosis treatment, for example.

POE and POU systems can be installed without smart controls, but the contribution of SWM technologies begins with the capability to monitor them. System monitors can be extended at the POE point or even past it to the feeder lines within the condominium complex. Communications with the monitors and controls can be wireless or wired. These can provide information on status, use and quality of water supplies, and this can be provided to water users to create transparency and foster trust in the systems. Other SWM tools that could be employed might be a dashboard to serve the full complex, the ability to move services from one user to another by rapid switching of services, and the ability to remotely shut off services to individual users. Barriers to such capabilities include security and the ease with which such controls might be bypassed or sabotaged, of course.

This arrangement offers the opportunity for a form of privatization, where a utility can offer concessions to manage the service extensions. Services could be from an established utility or through a new arrangement, which could be private, public or cooperative. Any of these would have to be regulated and have effective governance. Services could be at the 'basic' level and may require subsidy to motivate the utilities.

The example scenario shows the technical approach to use SWM extend services in one setting and how it can open new possibilities for business innovation with a public–private partnership. Innovations such as those shown for condominium water services are being tried in developing countries (Mara & Alabaster, 2008), and the new feature explained here would couple them with SWM technologies.

Conclusions

Smart water systems offer intriguing possibilities to improve public services, but how they can address access and affordability must be determined. Problems of access and affordability afflict all countries, though in different ways. In high-income countries access is high and the numbers of people with affordability issues are smaller than in low-income countries, but they remain significant because poverty is everywhere.

SWM systems are evolving, and it is too early to know their impacts, so the anticipated benefits are speculative. One benefit might be to use SWM to help people who lack service due to low-performing utilities or inability to pay. SWM can address the obstacles to these goals by improving utility operational results and converting low-income people to paying customers. Operational benefits include reducing water loss, improving hydraulic performance, detecting leaks and theft, and monitoring distribution systems, among others. This can facilitate customers to get access, increase policing and make the delivery more secure, legitimate and affordable. POU and POE systems can be employed in tandem with SWM to help with these goals, as well as with water safety.

Operational effectiveness is largely dependent on governance, infrastructure, equipment and workforce, which differ between high- and low-capacity utilities. Utility capacity for infrastructure and equipment is largely a function of community financial capacity, whether determined by local funding or transfers from higher levels of government. SWM might compensate for lack of financial capacity and strengthen utilities despite their limited resources. Governance and workforce are measured by human capacity and theoretically do not depend on financial capacity, as communities can organize effective governance or effective workforces even if financial capacity is lacking. This depends, of course, on community culture and capacity to address social issues, whether in cities or small communities.

The intriguing possibility is that if utilities can provide better services with more efficient systems, and more water users can become paying customers, access and affordability problems will diminish. To analyze this possibility, the status of housing can be studied in small communities and larger cities. By also considering the capacity of utilities serving these areas, three cases of interest were identified here: problems in small communities with low-capacity utilities, problems in cities with low-capacity utilities, and problems in cities with high-capacity utilities but unconnected housing.

When utilities have low capacity, SWM technologies might leapfrog over obstacles to gain control, reduce costs and extend service, provided that the means to implement them can be obtained. In a high-capacity utility, the problem may be financial constraints in serving low-income customers or lack of motivation. With SWM, the utility has new opportunities and motivation to identify illegal connections and enrol them as paying customers, even at basic levels.

In all cases, interventions to increase the effectiveness of operations address barriers such as lack of infrastructure, water loss, interruption of source of supply, and inadequate water treatment, among others. Interventions to improve the customer interface can help in identifying water users and illegal connections, as well as adding capacity to establish paying accounts.

One of the intriguing possibilities is that SWM systems can help in offering concessions to manage service extensions. The scenario shown as an example may be an idealized water supply system for the future, but it is based on current research and experience. In it, bulk water is distributed to zones of cities, which have their own storage and POE treatment systems.

Extending centralized services to the many people in less-than-conventional housing will be a slow process that is never complete. Centralized systems are expensive and capital-intensive. Alternatives to move from centralized to distributed services can help fill the gap.

Ultimately, good governance is required. No matter how innovative and entrepreneurial private operators are, access to water supplies in the commons, either surface or groundwater, must be regulated. Governance authorities must be involved somehow. Demonstration projects will be required, with adjustments of theories and methods that result from lessons learned.

Perhaps the largest concern about smart technologies is cybersecurity. Infrastructure systems can be sabotaged to create havoc and destruction. A related concern involves issues ranging from invasion of individual privacy, such as with unauthorized webcams, to hijacking of financial records and other crimes, that might be facilitated through the devices used in SWM.

Innovations such as those described in this article will not happen automatically, and there must be motivating political forces. With SWM to improve operations and deal with customers better, there is a real chance to improve access and affordability. Commitment by governance authorities at appropriate levels, financing, and demonstration projects will help advance the concept.

Disclosure statement

No potential conflict of interest was reported by the author.

References

American Water Works Association. (2016). *AWWA statements of policy on public water supply matters*. http://www.awwa.org/about-us/policy-statements.aspx

Black & Veatch Management Consulting. (2020). *50 largest cities rate survey: 2018-2019*. Retrieved June 4, 2020, from https://www.bv.com/sites/default/files/2019-10/50_Largest_Cities_Rate_Survey_2018_2019_Report.pdf

Cadmus Group. (2006). *Point-of-use or point-of-entry treatment options for small drinking water systems*. Report for US Environmental Protection Agency. https://www.epa.gov/sites/production/files/2015-09/documents/guide_smallsystems_pou-poe_june6-2006.pdf

Christophe, D. (2018). Smart water networks: Helping utilities keep the water flowing, or not. *Waterworld*, *34*(12), 14–15. https://digital.waterworld.com/waterworld/201812/MobilePagedReplica.action?pm=1&folio=14#pg16

Cohen, D. (2016). The rationed city: The politics of water, housing and land use in drought-parched São Paulo. *Public Culture*, *22*(2), 261–289. https://doi.org/10.1215/08992363-3427451

Cuppett, J., Clements, J., & Berahzer, S. (2016). *Affordability: A multi-faceted challenge for the water industry*. Water Research Foundation.

de Albuquerque, C. (2011). *Report of the Special Rapporteur on the human right to safe drinking water and sanitation*. United Nations General Assembly. http://www2.ohchr.org/english/bodies/hrcouncil/docs/18session/A-HRC-18-33-Add4_en.pdf

Drell, L. (2012). *25 technologies every smart city should have*. Mashable. https://mashable.com/2012/12/26/urban-tech-wish-list/

Eremia, M., Toma, L., & Sanduleac, M. (2017). The smart city concept in the 21st century. *Procedia Engineering, 181*(2017), 12–19. https://doi.org/10.1016/j.proeng.2017.02.357
Grigg, N. S. (2018). Service levels for the four billion people with piped water on premises. *Water International, 43*(4), 531–547. https://doi.org/10.1080/02508060.2018.1452119
Hopkins, O. (2018). *The challenge of reaching the poor.* AWWA Transformative Issues Symposium: Affordability. American Water Works Association. https://www.awwa.org/conferences-education/conferences/affordability.aspx
Hunter, P. R., MacDonald, A. M., & Carter, R. C. (2010). Water supply and health. *PLoS Medicine, 7*(11), e1000361. https://doi.org/10.1371/journal.pmed.1000361
Hutton, G., & Bartram, J. (2008). *Regional and global costs of attaining the water supply and sanitation target (target 10) of the Millennium Development Goals* (WHO/HSE/AMR/08/01). World Health Organization.
International Water Resources Association. (2018). *Smart water management: Case study report.* https://www.iwra.org/wp-content/uploads/2018/11/SWM-report-exec-summary-web-1.pdf
Joint Monitoring Programme. (2017). Progress on drinking water, sanitation and hygiene. https://www.who.int/water_sanitation_health/publications/jmp-2017/en/
Karwot, J., Kaźmierczak, J., Wyczolkowski, R., Paszkowski, W., & Przystałka, P. (2016). Smart water in smart city: A case study. Paper presented at *the SGEM 16th International Scientific Conference on Earth and Geosciences,* Albena, Bulgaria. https://doi.org/10.5593/sgem2016B31
Mara, D., & Alabaster, G. (2008). A new paradigm for low-cost urban water supplies and sanitation in developing countries. *Water Policy, 10*(2), 119–129. https://doi.org/10.2166/wp.2008.034
Melo, J. C. (2009). *Condominial water and sewerage systems.* World Bank. http://siteresources.worldbank.org/EXTWAT/Resources/4602122-1213366294492/5106220-1234469721549/32.1_Access_for_the_urban_poor.pdf
National Academy of Public Administration. (2017). *Developing a new framework for community affordability of clean water services.* https://www.napawash.org/studies/academy-studies/environmental-protection-agency
National League of Cities. (2016). *Trends in smart city development.* https://www.nlc.org/sites/default/files/2017-01/Trends%20in%20Smart%20City%20Development.pdf
Norton, J. W., Wood, B., Saparia, B., Jin, S., & Radtke, L. (2019). A journey into smart(er) waters: The Great Lakes Water Authority's experience and approach to smart water systems. *Advances in Water Research, 29*(4), 11–15.
Pereira, G., Parycek, P., Falco, E., & Kleinhans, R. (2018). Smart governance in the context of smart cities: A literature review. *Information Polity, 23*(2), 1–20. https://doi.org/10.3233/IP-170067
Phelps, G., & Crabtree, S. (2013). *Worldwide, median household income about $10,000.* Gallup. https://news.gallup.com/poll/166211/worldwide-median-household-income-000.aspx
Raucher, R., Hagenstad, M., Cotruvo, J., Narasimhan, R., Martin, K., Arora, H., Regunathan, R., Drago, J., & Pontius, F. (2004). *Conventional and unconventional approaches to water service provision.* AwwaRF Report.
Truelove, Y. (2018). Negotiating states of water: Producing illegibility, bureaucratic arbitrariness, and distributive injustices in Delhi. *Environment and Planning D: Society & Space, 36*(5), 949–967. https://doi.org/10.1177/0263775818759967
UN Department of Economic and Social Affairs. (2001). *Compendium of human settlement statistics 2001.* https://unstats.un.org/unsd/demographic/sconcerns/housing/publications/Series_N6.pdf
US Conference of Mayors. (2014). Public water cost per household: Assessing financial impacts of EPA affordability criteria in California cities. https://www.usmayors.org/pressreleases/uploads/2014/1202-report-watercostsCA.pdf
UN Habitat. (2016). *Slum almanac.* 2015-2016. https://unhabitat.org/slum-almanac-2015-2016/
Walton, B. (2016, March 22). *Water affordability is a new civil rights movement in the United States.* Circle of Blue. http://www.circleofblue.org/2016/water-policy-politics/water-rights-access/water-affordability-new-civil-rights-movement-united-states/

Institutional innovation and smart water management technologies in small-scale irrigation schemes in southern Africa

Henning Bjornlund, A. van Rooyen, J. Pittock, K. Parry, M. Moyo, M. Mdemu and W. de Sousa

ABSTRACT

This paper reports on the introduction of SWM technology, soil moisture and nutrient monitoring tools, alongside Agricultural Innovation Platforms (AIP) in three small-scale irrigation schemes in southern Africa. Quantitative and qualitative data are presented on the changes and benefits that have resulted, including increased yield and profitability. The findings emphasize that information prior and subsequent to adoption is needed, and the importance of understanding and enhancing the incentive framework for behavioural change, including both economic and physical returns. The findings illustrate SWM technology is strengthened when introduced with credible multi-stakeholder processes, such as an AIP, that facilitate institutional innovation.

Introduction

Water management is becoming more vital with the increasing withdrawal of water and competition over water resources, and the advance of climate change. Smart water management (SWM) technology can play a role in managing these complex water-related issues (Choi et al., 2016). Until now, SWM has typically been associated with supporting efficient water and wastewater management by industries and utilities in urban contexts at regional or city scales (for example, Ramos et al., 2020). SWM includes 'the use of Information and Communication Technology (ICT) to provide real-time, automated data for use in resolving water challenges' at a range of scales (Park et al., 2018, p. 25). The use of ICT for water management includes measuring rainfall, water quality and water flows; controlling treatment plants and sanitation systems; metering for billing customers; cloud-based data management; and efficient water management in urban contexts and irrigated agriculture (Kim, 2019; Park et al., 2018). This paper explores the importance of using innovative processes when introducing SWM technologies in small-scale irrigation schemes in southern Africa.

Irrigation accounts for more than 70% of water withdrawal globally (Wada et al., 2016), and the number of people subject to water scarcity is likely to quadruple by 2025 (Choi et al., 2016). Hence, there is an urgent need to improve water productivity in irrigation to help secure water supply for other users, without compromising food production. To do this irrigation needs to meet two challenges: produce more crops and use less water (Cai et al., 2011). Small-scale irrigation in southern Africa often involves less efficient furrow irrigation and, to manage risk, farmers tend to apply more water than is required (Moyo et al., 2020). To increase water productivity, irrigators need good knowledge of soil moisture dynamics (Stevens, 2006) to secure timely and efficient application of water to match a crop's water needs and prevent stress-induced losses in yield or quality (Wang et al., 2015). Whilst soil moisture monitoring tools offer the potential to improve irrigation efficiency and productivity, globally their adoption has been low, among other reasons due to high cost (Myeni et al., 2019; Stirzaker, 2006; Stevens et al., 2005).

Increasing irrigators' water productivity can, in addition to farmer-specific benefits, also have broader public benefits; such as meeting food sovereignty objectives and facilitating reallocation to other users whilst maintaining food production. SWM technologies might also be used to monitor and collect water management data from large numbers of farmers to support decision-making at various spatial scales. Achieving these benefits requires that many farmers adopt and use the technology in a sustained way (Park et al., 2018); hence, processes need to be put in place to overcome the barriers to adoption discussed in the following section of this paper. In particular, the technology used in Africa must be cost effective, simple to use, flexible, adaptive and provide clear financial and non-financial benefits to farmers (Akudugu et al., 2012; Corbeels et al., 2014; Foster & Rosenzweig, 2010; Mwangi & Kariuki, 2015; Pittock et al., 2020). According to Wang et al. (2015), the benefits for Canadian farmers include (in order of importance) reduced irrigation application, reduced labour and energy inputs, and increased crop yields and quality. This has synergies with the advantages perceived by African farmers, which include (in order of relative advantage) water saving, fuel saving, labour saving and yield increase (Adimassu et al., 2020). It is important to note the multiple economic benefits expressed by farmers in addition to the physical yield return. As recently argued by Michler et al. (2018), belief in the positive association of yield increases (in particular) and economic return has confounded the adoption literature as there are other factors such as markets that influence economic return.

Considering that improved water productivity has public benefits and cost is one of the identified barriers to adoption, this raises several questions: who should pay for the technology; how much are farmers willing to pay to gain the personal benefits; and how much should the public pay to ensure sufficient adoption to secure the public benefits? Cost-sharing arrangements to achieve public and private benefit in irrigation already exist: for example, in Alberta, Canada, the mixture of public and private benefits from investments in irrigation has resulted in a sharing of the cost of maintenance of irrigation infrastructure between irrigators and taxpayers (Bjornlund & Klein, 2015). With respect to technology, recent analyzes of small-scale irrigator's willingness to pay for soil moisture monitoring tools in Africa found that while farmers indicate a willingness to pay, there would still be a need for some co-investment by other public organizations to ensure widespread adoption (Abebe et al., 2020; Adimassu et al., 2020).

This paper reports on a project implemented in southern Africa called 'Increasing irrigation water productivity in Mozambique, Tanzania and Zimbabwe through on-farm monitoring, adaptive management and agricultural innovation platforms' (2013–17), subsequently renamed 'Transforming small-scale irrigation in southern Africa' after the second phase (2017–21) was funded. Hereafter, called TISA (2013–17) or TISA (2017–21) when referring to the separate stages of the project. TISA illustrates the need for multiple entry points to improve yield and profitability. The entry points included the provision of SWM technologies, in the form of two types of soil moisture and nutrient monitoring tools (the tools), and Agricultural Innovation Platforms (AIPs) (Figure 1). Together, these complimentary entry points were expected to encourage a transitional and self-sustaining process of economic growth and community development (Pittock et al., 2020; Van Rooyen et al., 2020). When developing the TISA project, it was recognized that AIP processes were required to stimulate learning, feedback and institutional innovation at critical points in the system to overcome the socio-institutional barriers to adoption and ensure that increased production resulted in increased profitability. Institutions in the context of this paper are distinct from organizations and are the formal and informal rules that influence interactions and decision-making, encompassing legislation, property rights, traditions, codes of conduct, and norms (North, 1991; Ostrom, 2005).

This paper analyzes the process of introducing the AIPs and tools into three small-scale irrigation schemes in Mozambique, Tanzania and Zimbabwe since 2013 (Table 1).

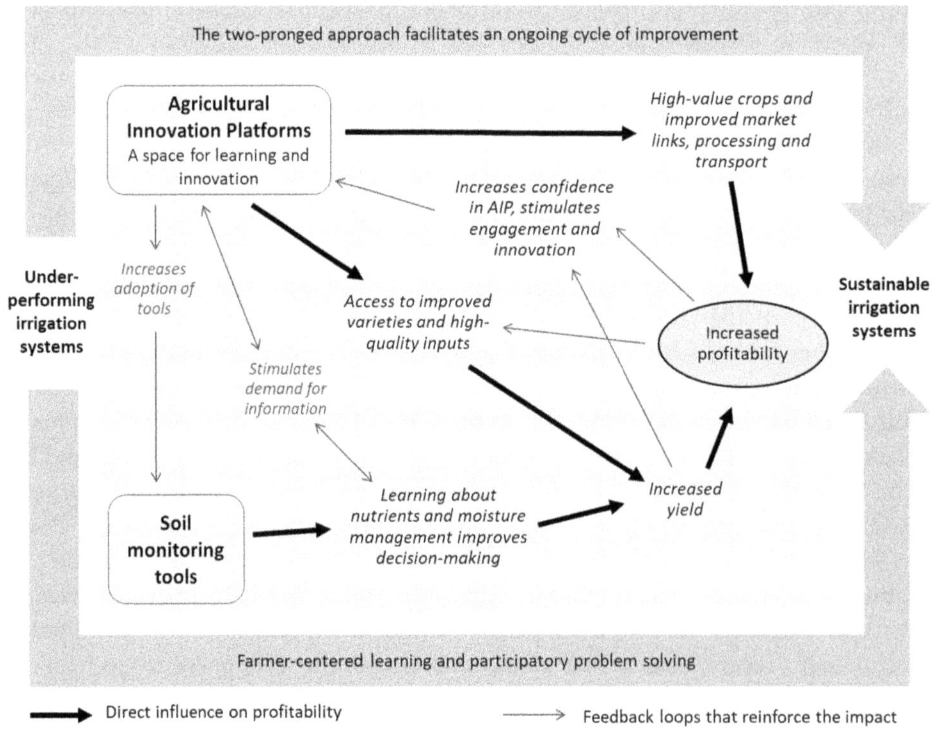

Figure 1. The two-pronged approach (Bjornlund et al., 2018).

Table 1. Characteristics of the small-holder communal irrigation schemes at project inception.

	Tanzania	Mozambique	Zimbabwe
	Kiwere	**25 de Setembro**	**Silalatshani**[2]
Year constructed	2004–07	1975	1968–69
Location	Iringa District	Boane District	Insiza District
Farmers	168	40	845
Irrigated (ha)	195	38	442
Main crops[4]	Tomatoes, onions, green maize	Cabbages, tomatoes, green beans	Maize, wheat, sugar beans
Legal structure	By-laws	By-laws	By-laws
Land access	Inheritance, buying, renting[1]	Cooperative holds land use title	State-owned, chief allocates
Soils	Sand & clay of varying fertility	Mostly fertile clay soils	Mostly loamy sandy soils
Rainfall (mm)	700	650–900	450–650
Main bodies governing water	Basin Water Board	Regional water authorities	Catchment Councils
Irrigation water source and conveyance method	River, gravity canal	River, motor pump	Dam, gravity canal
Irrigation method	Surface flooding	Surface flooding	Surface flooding

Sources: project data. [1] In Tanzania, farmers have the legal right to private ownership but very few do so due to transaction costs. [2] At Silalatshani the project was conducted within the Landela Block, which has 212 farmers. In this paper when reference is made to Silalatshani, it refers to the Landela Block.

The introduction of the tools was coordinated by three TISA country teams, comprising local researchers and agricultural development practitioners. The paper presents some of the early signs of success and lessons from the process of introducing SWM tools in conjunction with an AIP. Individual and social learning has taken place, with farmers creating, exchanging and spreading new irrigation knowledge across the schemes (Parry et al., 2020). As a result, both the frequency and duration of irrigation have been reduced, soil nutrients have been retained within the root zone, and yields and income have increased. These and other impacts from the project are discussed in more depth elsewhere (Chilundo et al., 2020; Mdemu et al., 2020; Moyo et al., 2020; Parry et al., 2020; Stirzaker et al., 2017; Van Rooyen et al., 2020). This paper focuses on the processes used to implement the two-pronged approach, and we argue that the positive results have been enhanced by the synergies between farmers learning from the tools and the AIP generating incentives for critical changes through institutional innovation.

The first section discusses the factors influencing technology adoption. This is followed by two sections describing the AIP process and the SWM tools, respectively, how they work and the process of implementation. This is followed by a description of the synergies of the tools and AIPs, and how their interactions improve farmers' livelihoods and, hence, provide the incentive to use the tools and collect the data. Two further sections briefly discuss how the tools were modified to automate the consolidation of the individual data into a large database, and how to out- and up-scale the two-pronged approach.

Adoption of SWM technologies, findings from the literature

There is a wealth of literature identifying the factors that influence the adoption of agricultural technologies (Mwangi & Kariuki, 2015). Many studies focus on the characteristics of farm households (e.g., gender, household size, wealth, access to credit, health, education, learning and experimentation, level of dependency on farming), and characteristics (e.g., farm size, soil quality, crop choices, and access to advice) (Abdulai & Huffman, 2014; Akudugu et al., 2012; Corbeels et al., 2014; Foster & Rosenzweig, 2010; Wang et al., 2015; Wheeler et al., 2017). These studies highlight the diversity of farm households and farming conditions and suggest that constraints to adoption cannot be generalized across locations; reflecting this, technology that is profitable in one location may not be profitable elsewhere (Foster & Rosenzweig, 2010).

Fundamental to farmers' adoption is what they know about a technology, how easy it is to use, how useful and beneficial it will be to their production, and whether it supports increased profitability. The provision of information on the technical aspects only, or the intended water conservation benefits of the technology, will be counterproductive in all but the most straightforward of settings (Glover et al., 2019; Perret & Stevens, 2006). Rather, the technologies should be tailored to be easy to use and recognize small-scale farmers' objectives, which are often focussed on short-term production for survival and providing for the family (Corbeels et al., 2014; Perret & Stevens, 2006). Farmers also need to consider the changes they may have to make when adopting the technology. Households on small-scale irrigation schemes typically manage their resource allocation across multiple livelihood strategies, encompassing irrigated and dryland farming, livestock and off-farm activities (Bjornlund et al., 2019). Hence, adoption of water conservation technologies that intensify production includes decision-making across economic activities; whereby resources allocated to one activity may impact another (Corbeels et al., 2014). Ultimately, 'money matters' and economic measures of return should be factored into the influences on the adoption of agricultural technologies (Michler et al., 2018).

In terms of the adoption process, Perret and Stevens (2006) assert that integration of a technology and innovation is fostered through farmers' participation in technology development so that it meets their needs and expectations and responds to actual demand. This favours a bottom-up approach. Farmers' decisions to adopt an irrigation technology are influenced by obtaining information on the technology from other farmers and extension agents (Wang et al., 2016) and the earlier this takes place the better the outcomes (Foster & Rosenzweig, 2010). Neighbourhood effects have been identified on irrigators' adoption of water markets (Haensch et al., 2019), and the learning from the use of soil monitoring tools (Abebe et al., 2020). The intensity of adoption is increased by farmers continuing to receive advice following adoption (Wang et al., 2016). In this regard, the provision of information about the benefits, support services post-adoption, peer groups to disseminate experiences and business advice will improve the outcome of adoption (Wang et al., 2016).

In government irrigation schemes, decision-making encompasses a diversity of interacting actors: for example, actors in schemes in Zimbabwe include farmers, irrigation management committees, government agencies, extension services and the private sector (Van Rooyen et al., 2020). Hence, schemes are now understood as complex systems, requiring collective action for water management (Ostrom, 2007; Perret & Stevens, 2006). These complexities influence the benefits farmers can derive from adopting SWM technologies. Perret and Stevens's (2006) model of the conditions influencing adoption reflects the complexity and highlights two issues: i) the need for interactions between water conservation technology, collective action and institutions; and ii) that these components need to be sound, sustaining and enabling, in order for adoption to be successful and result in sustainable and beneficial outcomes (Figure 2). The degree of collective action required increases as the spatial scale of water management increases (Perret & Stevens, 2006).

The model is consistent with the notion that linear technology transfer and diffusion approaches, from scientist to extension agents to farmers for adoption, are inadequate (Glover et al., 2019; Knickel et al., 2008). Instead, the introduction of technologies should be part of an innovation system; whereby, learning and innovation involve collaboration across a range of knowledge producers (Maru, 2018). This accords strongly with the rationale of the TISA project. In complex socio-economic contexts, such as small-scale irrigation schemes, this requires that a technology can be re-innovated, co-constructed and translated for use in its new setting (Garb & Friedlander, 2014). Therefore, technology adoption must include farmers adapting and using technologies to suit their farming activities and aspirations, which may include changing their farming practices (Corbeels et al., 2014; Perret & Stevens, 2006).

Currently, small-scale irrigation schemes in SSA experience many challenges, and many are performing poorly with low productivity at scheme and farmer scale (Mutiro & Lautze, 2015; Stirzaker & Pittock, 2014; Sullivan & Pittock, 2014). Consequently, the enabling environment (Figure 2) is not conducive for the collective action required to facilitate the transition processes and successfully introduce a new technology. An institutional issue that is particularly pertinent is that irrigators do not contribute to decisions regarding irrigation scheduling. For example, water is supplied through fixed infrastructure and irrigation schedules are often enforced by an irrigation management committee (Moyo et al., 2020). Market conditions, input support and subsidies, agricultural policies, availability of infrastructure and property rights are other constraints

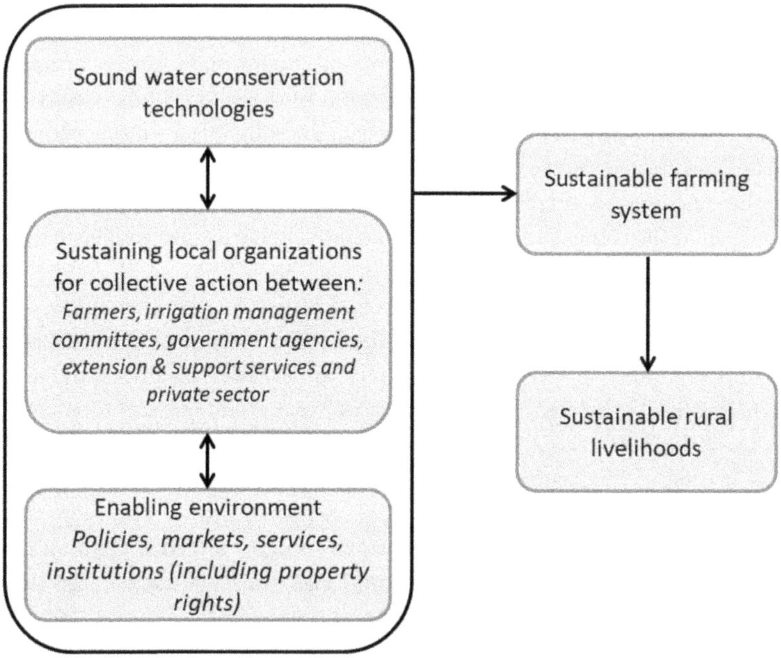

Figure 2. Conditions for water conservation technology adoption towards sustainable farming systems and rural livelihoods (Source: adapted from Perret & Stevens, 2006).

(Bjornlund et al., 2017); an understanding of these conditions is a pre-requisite for technology introduction (Andersson & d'Souza, 2014; Corbeels et al., 2014; Perret & Stevens, 2006). Affirming the importance of markets, Michler et al. (2018) assert that differing adoption rates may be explained by the existence or functionality of markets. Improving water use productivity, therefore, sits in the context of many other barriers to profitability, and institutional innovation is required alongside the introduction of SWM technology. Woltering et al. (2019) argue that overcoming barriers requires development actors to have an understanding of multiple overlapping systems, a clear picture of the elements of the system that the actors cannot solve, and where they can have an influence through strategic collaboration to kick-start the process of change. Sustainable system change is evidenced when institutions and incentives are such that the system achieves its objectives. To achieve systemic change, learning and experimentation need to reach all stakeholders in the system (Independent Science and Partnership Council [ISPC], 2015).

Agricultural innovation platforms[1]

AIPs in complex systems

Multi-stakeholder platforms such as AIPs have been widely used in a diversity of farming contexts in Africa such as seed, maize, honey and livestock production (Makini et al., 2013), and are used by TISA for the first time within the target irrigation schemes. AIPs are particularly suited to address multiple constraints in complex systems such as irrigation schemes (Van Rooyen et al., 2017b). Complexity arises due to the range of sub-systems, associated actors, and the multiplicity of interactions (economic and otherwise) that comprise the system (Figure 3). Complexity is further compounded by the increased risk and wider skill set associated with irrigation compared to dryland farming. Ideally, well-functioning and adaptive irrigation systems function with decentralized control. However, in many government-managed irrigation schemes in SSA, irrigation management bodies control key functions such as irrigation schedules and cropping calendars with little feedback from farmers. An inherent feature of complex systems are the information flows and feedback mechanisms that determine the outcomes of interactions between stakeholders, which can provide incentives for change, facilitate learning, and increase the adaptive capacity and resilience of the system (Abson et al., 2017; Meadows, 2008).

An inherent feature of innovation platforms is that they support the diffusion of innovation, by providing a participatory and decentralized innovation space to identify solutions to system barriers (Pamuk et al., 2014). They are increasingly used in agricultural research-for-development contexts, where a research or development entity initiates their establishment (Pamuk et al., 2014; Schut et al., 2017). By facilitating interaction between stakeholders, an AIP acts as a catalyst to create an informal network that assists with improved system functioning (Figure 4). Thus, the value of an AIP in the context of technology adoption, in systems lacking functionality and an enabling environment, is readily apparent: that is, the AIP should assist in overcoming the socio-institutional and economic barriers that limit the profitability a technology may afford at farmer and scheme scale.

In the TISA context, the AIP was introduced to facilitate informal networks of stakeholders and overcome many of the systemic barriers to adoption identified in the literature review (for example, Andersson & d'Souza, 2014). A facilitator is used to

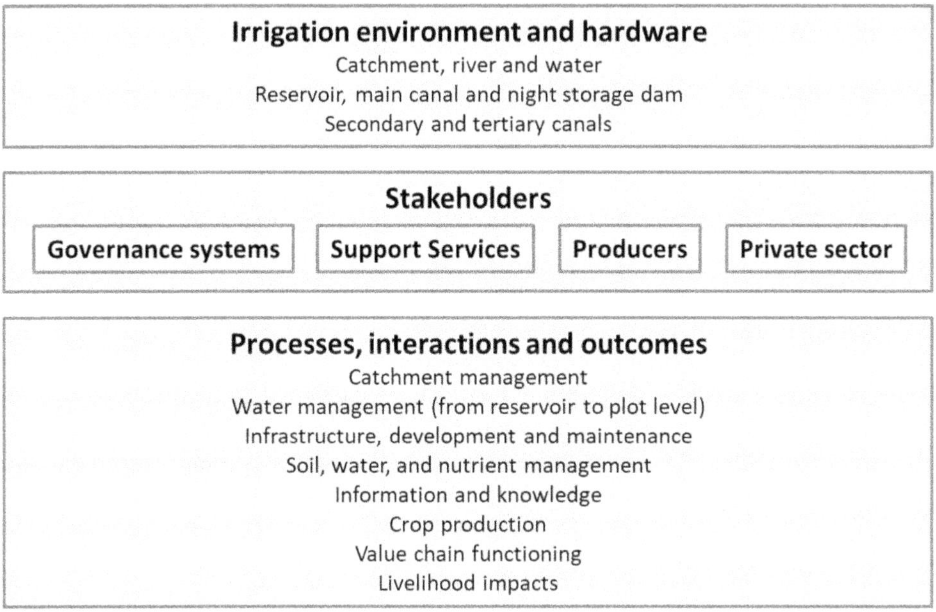

Figure 3. Example of a Zimbabwe irrigation system and its components (adapted from Van Rooyen & Moyo, 2017a).

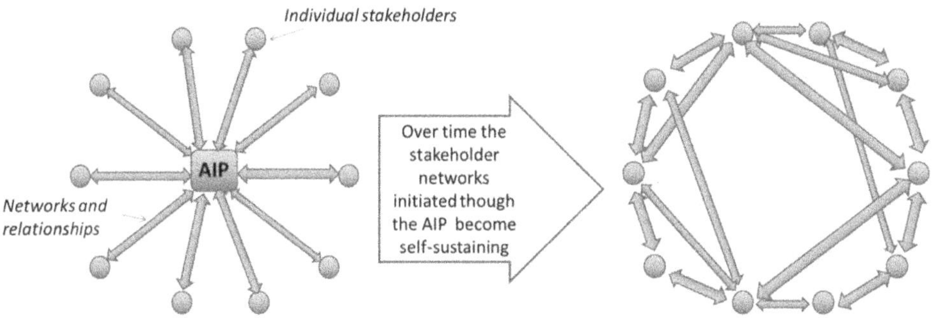

Figure 4. The role of AIPs in developing functional stakeholder networks (Van Rooyen & Moyo, 2017a).

maintain a coordinating role, monitors progress and maintains transparency of the process. Over time, the relationships and networks of an AIP may be formalized such as in public–private partnerships (Schut et al., 2017). In the long term, the AIP may become self-sustaining, it may evolve to focus on new challenges with a potentially new mix of stakeholders, or it may no longer be required (Schut et al., 2017; Van Rooyen et al., 2017b).

Setting up an AIP

The three TISA country teams undertook AIP training to understand and lead the establishment of the AIP process. Information on each scheme (Table 1) was collected to better understand the context: agricultural system, demographics, irrigation scheme management and the challenges faced by each scheme (De Sousa et al., 2017; Mdemu et al., 2017; Moyo et al., 2017). This information was used in conjunction with local experts to identify an initial list of AIP stakeholders (See Box 1 for initial expert engagement leading to identification of the AIP members). Stakeholders are context-dependent, but the aim was to have representation across those who have an interest in the performance of the irrigation scheme, including from across the value chain. Box 1 provides an example of stakeholders from the Silalatshani AIP, illustrating that a diversity of stakeholders are engaged from the system: governance, support services, irrigators and the private sector. In identifying potential stakeholders, the following characteristics were considered: understanding of the scheme's challenges; ability to identify options to address the challenges; being innovative and active in the system; and disseminators of knowledge and information.

The success of AIPs is underpinned by skilled facilitators and the networks and connections that are established through the AIP. The facilitator's role is to create a space for interaction and networking, analysis, sharing of information, experimentation and learning, and to foster ownership of the forum by the stakeholders. The facilitator is critical in guiding the stakeholders through the early processes by encouraging and entertaining all suggestions; maintaining transparency of the process by keeping stakeholders informed of decisions and progress; supporting participation by all AIP stakeholders; and accommodating failure as a normal part of learning (Van Rooyen et al., 2013). Three facilitation options were implemented: led by a development organization with no affiliation to any of the stakeholders involved (Tanzania); government-led (Mozambique) and researcher-led (Zimbabwe) (Pittock & Stirzaker, 2014).

The four-stage AIP process

Once the facilitator and main stakeholders are identified, an AIP commonly progresses through four stages. In the TISA project, Stages 1, 2 and 3 were undertaken in a 2-day workshop in each scheme. For a comprehensive discussion of these processes in TISA see Van Rooyen et al. (2017b).

Stage 1: introducing the AIP

Box 1. An example of stakeholder engagement and selection from the Silalatshani AIP.

Initial engagement took place at a meeting with the District Administrator for the Insiza District in October 2013. Also present were: i) District Extension Officer, Extension Officer Agronomy, and Silalatshani Extension Supervisor from Ministry of Agriculture/Provincial Offices (AGRITEX); ii) district irrigation technician from Department of Irrigation (DoIRR); and iii) the project country co-ordinator and scientist from the International Crops Research Institute for the Semi-Arid Tropics (ICRISAT).

The stakeholders identified to participate in the AIP were:

(i) Governance systems: a) Head Offices of the Ministry of Agriculture; Mechanization and Irrigation Development; Ministry of Environment, Water and Climate; Zimbabwe National Water Authority (ZINWA); AGRITEX; DoIRR and Department of Mechanization; b) District Administrator; local Chief; Department of Livestock Production; Ministry of Lands & Rural Resettlement; District Development Fund (DDF); Ministry of Social Welfare and the Police from Local Government; c) representatives of Silalatshani's Irrigation Management Committee (IMC)

(ii) Support Services: a) District government officials from Department of Public Works; Zimbabwe National Water Authority (ZINWA); Ministry of Youth Development; extension staff (AGRITEX); DoIRR provincial staff; and Insiza Rural District Council; b) NGOs (ZIMAED and Bulawayo Projects Centre).

(iii) Producers: irrigators from Silalatshani

(iv) Private sector: financiers from AGRIBANK; value chain stakeholders such as agri-dealers

The AIP process is explained, the benefits are described, and an agenda and process are agreed. Participants are asked to explain their interest, confirm their role and commitment to the process, and identify additional important stakeholders. In Silalatshani, the stakeholders described the initial workshop as a major breakthrough as they had never had the chance to meet and discuss the common issues. There was collective agreement of the need to work together. For some key stakeholders – for example, the Zimbabwe National Water Authority (ZINWA) – there was recognition of their critical influence on scheme production.

Stage 2: identification of system constraints

This stage allows for full disclosure of problems and a thorough discussion of root causes. First, participants work in separate groups (e.g., farmers, technical support staff, private sector and government) to list and prioritize constraints. The facilitator ensures there is adequate time for all significant challenges to 'surface', as inadequate time spent on this step can restrict individuals' ability to move on to subsequent stages. In order to think critically about the constraints, the facilitator asks participants to repeatedly consider the 'why' question to identify the root causes of the constraint and, hence, potential solutions. In this way, the stakeholders move beyond a generic articulation of a challenge and its cause. Once the challenges are clearly listed and their interactions clarified, the groups brainstorm potential solutions and the stakeholders who can implement them (Table 2).

In Silalatshani, this stage confirmed the need to initiate activities to address the constraints, including the important role that the soil monitoring tools could play in this context. For Kiwere, an important outcome was the confidence instilled in the stakeholders, and with this positive change in mindset they recognized their collective ability to address the scheme's challenges. Through this stage, participants also gain an understanding of the interconnected nature of the problems, and the feedback mechanisms that are operating. For example, limited access to inputs results in poor yields and, subsequently, low returns. Similarly, poor markets lead to low income and reduce the incentive and capacity to invest in inputs.

Table 2. Examples of initial identification of system constraints (Van Rooyen et al., 2017b).

Scheme	Systems constraints
Silalatshani (Zimbabwe)	Water payment arrears to ZINWA Infrastructure maintenance (fences and canals) Crop selection Farmer capacity Water management and allocation
Kiwere (Tanzania)	Poor infrastructure (canals not lined) Outdated agricultural equipment High price of inputs (fertilizers are tampered with) No reliable markets
25 de Setembro (Mozambique)	High pumping costs Poor infrastructure (canals not lined) No reliable markets (transport network is poor, leading to high transport and input costs) Farmer capacity (lack of knowledge and extension officers largely absent) Access to finance and lack of inputs

Stage 3: visioning

Participants work in their groups to develop a picture of the current layout and features of the scheme and their vision for the scheme (Figure 5). The facilitator motivates participants to 'dream' for an improved future and the pictures stimulate this. Importantly, the vision for the scheme should express what stakeholders perceive to be achievable within five years but not restricted by whether the pathway to attain this is clear.

The participants also develop a narrative of how their scheme might move from the present to the future situation (Box 2). In this way, farmers' aspirations and the context of interventions can be communicated to other AIP stakeholders who can then ensure that their technological and policy interventions link with the community's vision.

It is not uncommon that infrastructure issues are identified as the root cause of the problems with hard engineering interventions suggested as quick fixes: however, this is despite the documented failure of infrastructure improvements to improve scheme functionality (Fanadzo et al., 2010). Many projects, therefore, finish the diagnostic process with only a partial understanding of root causes; however, the AIP visioning stage encourages participants to think more deeply about the barriers and enables the socio-institutional and broader system issues to emerge. For example, while yields may be increased through improved agronomic practices, this will not necessarily result in increased income unless the constraints of storage, transportation and poorly functioning markets are addressed.

Stage 4: innovation process

The linkages established through the AIP process and across the system's stakeholders become paramount in the innovation stage, as many constraints are associated with

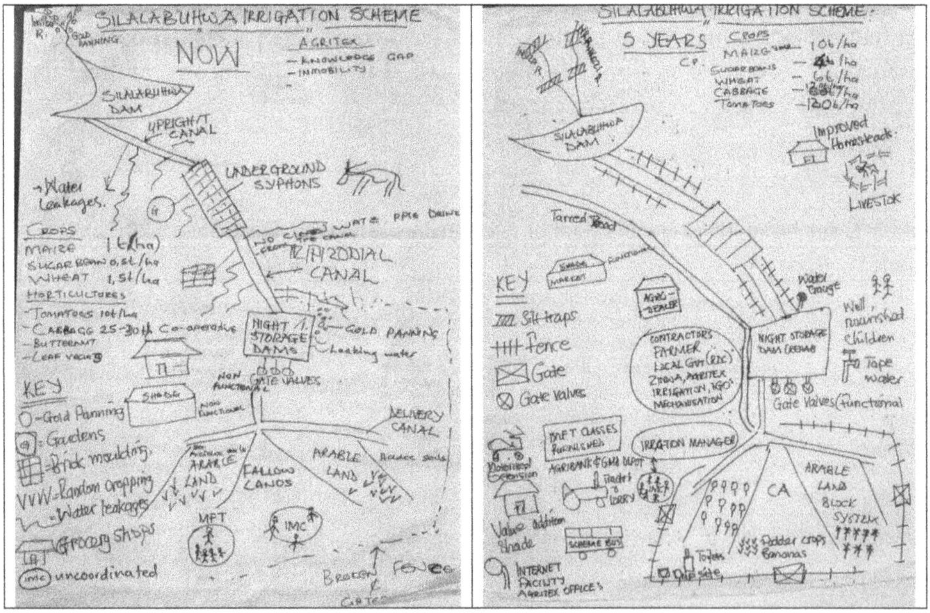

Figure 5. Current situation (left) and desired situation (right) for Silalatshani, Zimbabwe (Van Rooyen et al., 2017b).

Box 2. Present state and desired future narratives from Silalatshani.

Present state of the scheme

The cropping programme from the 1960s, commanding the production of low value staple crops, is still followed even though farmers cannot make a profit. Yields are low, often similar to rain-fed production. Post-harvest infrastructure is non-functional, with sheds and grading infrastructure damaged. There are disputes over unpaid water bills to ZINWA; many farmers have opted out of the scheme increasing the burden of water charges and maintenance on the remaining irrigators. There are no agreed structures to transfer unproductive land to willing farmers. AGRITEX staff lack confidence to promote new techniques and higher value crops and cannot access training courses to improve their knowledge. There is poaching of water from the canal between the dam and scheme, which is used for brick-making, livestock and home gardens but members of the scheme pay for the water. The IMC is largely defunct and does not have a voice in the running of the scheme. Failing infrastructure leads to water losses and broken fences mean livestock stray onto the scheme. Land degradation is evident, with areas of low pH or high water tables. Illegal gold panning contributes to land degradation and damaged canals.

Envisaged state of the scheme in five years

The unpaid water charges have been resolved. The scheme is more productive, with maize yield rising from the current 1 t/ha to over 4 t/ha. Diversification to high value crops such as garlic and other vegetables with a coordinated marketing strategy. Farmers are linked to financial institutions to access credit for purchasing inputs. Farmers invest in more inputs and water saving techniques. Currently unused land is in production and profitably used to grow fodder to fatten livestock. There is improved mechanization and infrastructure such as roads, fences and canals are repaired. There is post-harvest storage and processing and improved conservation measures, with silt traps and catchment management plans.

policy, infrastructure, and markets. Hence, the value of including influential stakeholders who come to recognize their role through the process: for example, microfinance and commercial banks, District Services for Economic Activities (SDAE) of Boane and the National Institute of Irrigation (INIR) in the 25 de Setembro AIP (Mozambique). There are potentially many challenges to address; hence, stakeholders are asked to identify the most critical barriers, potential realistic solutions, and the sequence of activities needed to implement the solutions. Importantly, shared ownership is established for a prioritized set of solutions for the most critical constraints (Table 3). The AIP organizes the most appropriate stakeholders into sub-groups to implement, evaluate and refine solutions.

As the formal problem identification processes are often focussed on agricultural productivity issues, there may be some 'deal breaker' issues that emerge either during the AIP meetings or at a later stage. There must be clear attempts to resolve these issues otherwise trust and commitment by stakeholders may be compromised. For example, at Silalatshani the ZINWA bill and unfarmed plots resulted in low system productivity and very few irrigators paying their water bill. Land is a sensitive issue in Zimbabwe and beyond the immediate scope of the project to effect change; however, after extensive discussion during the AIP meeting a team led by the Insiza District Administrator and the local Chief decided they would conduct a land audit to resolve the issue of absentee landowners (Table 3).

The constraints listed in Table 3 represent entry points for action and change in the system. However, there must be an incentive for the actors to make a change. In the case of the unpaid ZINWA water bill, the incentive became clear when the stakeholders discussed the issue: without a resolution, farmers would not resume farming, the scheme would collapse and ZINWA would not receive any money. Importantly, some actions were initiated outside the AIP process. In the Silalatshani visioning exercise, farmers expressed the desire for improved fencing (see Figure 5 and Box 2), but this was not of sufficient priority for further action to be initiated through the AIP process. However, farmers voluntarily paid extra money to the Irrigation Management Committee to have

Table 3. Examples of identified constraints, actions taken and impact at three irrigation schemes.

Constraint	Action taken and who initiated it	Impact of the action
Ageing tractor and truck (deS)	Contact was made with a Japanese NGO, which had previously assisted.	New truck and tractor acquired.
Water debt of US$286,000 owed to ZINWA (S)	AIP brought the issue to the surface, and ZINWA recognized its importance. They engaged with the IMC to discuss its resolution.	ZINWA recalculated the outstanding debts and interest, reducing the debt to US $80,000 at 1% interest per annum to be paid off as soon as possible.
Insufficient irrigation water and high irrigation costs (deS)	INIR worked with the farmers to design an irrigation rehabilitation strategy.	Infrastructure was rehabilitated. Speed of supply increased lowering time spent irrigating and diesel cost.
Poor integration into input and output markets (S)	Linking farmers to an existing produce buyers' forum in Bulawayo, organized by Bulawayo Projects Centre, a local NGO and AIP participant.	Readily available inputs, improved quality of input, new varieties and higher value crops, leading to improved yields, sales, gross margins and farm incomes.
Bogus suppliers selling poor quality seed and fertilizer (K)	AIP negotiated linkages to credible suppliers and organized bulk purchases. Main input suppliers provided farm input demonstrations.	Access to quality inputs without transport and handling costs. Improved farmer knowledge on input use and management. Increased yield and farm income.
Low farm income (deS, S)	Tools installed, and farmers connected to markets (S).	Increased yield converted to increased income.
	Workshop with farmers computing gross-margins for key crops (deS).	Better crop selection, increased income.
Absentee landowners and unfarmed plots. (S, deS)	District leadership and AGRITEX initiated a plot audit. Unfarmed plot owners encouraged to use their plots (S).	Improved land use.
	Farmers and extension officers discussed how to use unoccupied plots and engage young farmers. Absentee landowners told they would lose their plot if not used (deS).	Unfarmed land is used by other family members or reallocated to young farmers. Elderly farmers are renting their plots to other farmers.
Conflicts over boundaries, plot sizes and water fees (K)	Participatory mapping to show agreed boundaries and plot sizes. Mapping also used as an information system.	Common agreement on boundaries. Improved collection of water fees. Reduction in boundary conflicts. Improved enforcement and accountability.
Inefficient management of irrigation water and high in-field water losses (S, K)	Introduction of soil moisture and nutrient monitoring tools and training on their use (S, K).	Widespread adoption of the learnings from the tools improved water management, reduced water losses, saved time, increased yield and improved supply to downstream users. Reduced conflict over water access.
Lack of agronomic knowledge and low productivity (S, K, deS)	ICRISAT and AGRITEX initiated: i) paired demonstration plots; ii) training workshops on improved agronomic practices; iii) exchange visits to other schemes; and iv) on the spot training on specific issues (S).	Agronomic knowledge enhanced and improved crop yields and incomes from the crops.
	Visit to Igomelo scheme by farmers, extension, agricultural and project officers. Linkage to development organization providing training on agronomic practices (K).	Adoption of good agronomic practice. Use of organic manure. Improved productivity and profitability. Revision of the scheme's constitution.
	Field trip and training of farmers and extension officers (deS).	Tomato yield and income increased.
Unknown fertilizer application rates (K)	Soil fertility analysis provided fertilizer recommendations for different crops on different soils.	Improved soil fertility lead to improved crop yield and income.
Poor crop selection (S)	Training of support services in gross margin analysis and linking farmers to better markets. Initiated by AGRITEX and ICRISAT.	The IMC relaxed the cropping calendar and farmers improved their crop selection. Improved profitability.

(*Continued*)

Table 3. (Continued).

Constraint	Action taken and who initiated it	Impact of the action
Invasion of cultivated fields by cattle (deS)	Meeting with livestock farmers to discuss the issue of invasion. Identification of a new pathway for cattle to access water and ways for farmers to save money to build fence around the fields. Initiated by SDAE and INIR	Fence contracted and crops damage by livestock reduced. Improved yield and productivity of crops.
Lack of irrigation maintenance program (deS)	Workshop with farmers and extension officers to develop business plan for scheme to assess the cost of system and building maintenance. Crop preferences, agronomic practices and yields were discussed.	Business plans for scheme maintenance and for three crops (chilli, green maize and tomato).

S, K and deS denote Silalatshani, Kiwere and 25 de Setembro, respectively.

the fence built; demonstrating their willingness to participate in collective action and their capacity to self-organize and progress issues.

Soil monitoring tools

Background and theory

Climate, crop and soil data are traditionally considered the best information to decide when and how much water to apply to a specific crop (Stirzaker et al., 2017). This method has several weaknesses, as each type of data is prone to variability: climate is variable; crop choice and planting density can change; and application of precise volumes of water is not possible. Further, access to this data is difficult and most farmers do not have the skills to convert it into informed irrigation decisions. Hence, farmers generally rely on instinct, knowledge, experience and confidence gained over many years (Stevens et al., 2005), rather than objective measures. The literature emphasizes the challenges associated with the adoption of technology including the cost-effectiveness and benefits accruing to farmers; the information provided to farmers about the technology; the need for farmers to adapt technology to their context; and the critique of the linear technology adoption model (for example, Corbeels et al., 2014; Foster & Rosenzweig, 2010; Glover et al., 2019; Perret & Stevens, 2006; Pittock et al., 2020).

In developing new soil moisture monitoring technologies for southern Africa, consideration was given to the different mental models used by scientists and farmers, the specifics of the data that would be useful for irrigators and how the technology could work as a learning system in the local context (for more detail see Stirzaker et al., 2017). This approach is in contrast to earlier practices of top-down information provision (Pittock et al., 2020). The soil monitoring tools used by TISA draw on the theory that information systems for small-scale irrigators in developing countries should:

- offer people-centred and experiential learning;
- promote adaptive learning and management by integrating experience, observation, monitoring and feedback; and

- be inexpensive, robust and suitable for farmers with low literacy and numeracy: simple to use and provide the least amount of information needed for irrigation decision-making (Stirzaker et al., 2017).

The tool's designers drew on the science of soil water and solute measurement techniques to determine the parameters that should be measured. Soil moisture is measured through soil tension, which is a measure of the force a plant needs to use to extract moisture from the soil and avoids the need for calibration for different soil types. Nutrient status is assessed by measuring nitrate, as this is particularly susceptible to leaching and is the main form in which soluble nitrogen is available to plants. The final parameter is salt, as salinity leads to land degradation and reduced yield. It is important that the parameters are measured together, so farmers understand the relationships between them and what is happening in the soil: hence, they can make more informed decisions about the scheduling of water and fertilizer application (Stirzaker et al., 2017).

Description of the soil monitoring tools

Farmers received two devices:

- Chameleon™ Soil Water Sensor array and reader (Figures 6 and 7); and
- FullStop™ Wetting Front Detector (Figure 8) along with an electrical conductivity metre and nitrate test strips.

The Chameleon was first installed through TISA at Kiwere in mid-2014 and measures soil tension using an inexpensive resistance-type sensor. Three sensors (initially four) are combined into one array and permanently buried in the soil, so that soil moisture can be measured in the top, middle and bottom of the root zone. The sensor array can be

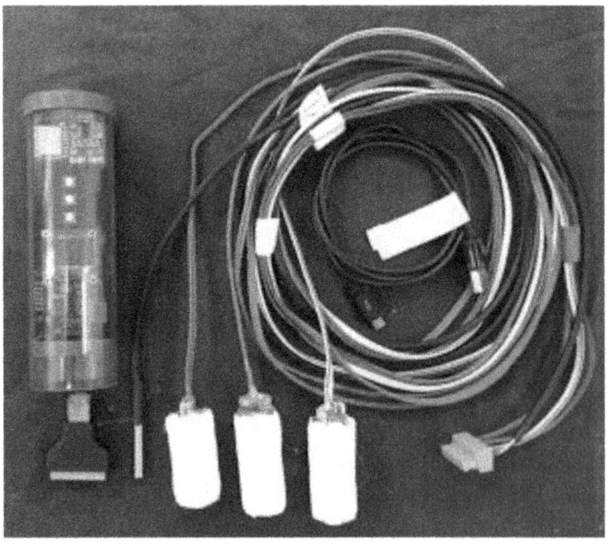

Figure 6. Chameleon™ soil moisture sensors and reader (VIA, 2019).

Figure 7. Farmer demonstrating the use of the Chameleon reader at Kiwere scheme (Photo: Karen Parry).

plugged into a reader, which is shared amongst farmers. The reader has a light for each of the sensors, which changes colour to blue, green or red according to the soil moisture at each depth:

- blue, soil is wet;
- green, soil is moist;
- red, soil is dry.

These features make it easy for the farmers to use as the colours provide a common language about soil moisture, which is independent of soil type, salinity level and temperature. While soil type does not influence the reading of red, green or blue, it influences how quickly the colour change occurs as the material in the sensor dries out. These features catalyse ongoing farmer observation and recording of the data is critical to their circumstances, and farmers learn when to irrigate in response to the colour pattern.

The Wetting Front Detector was developed in the early 2000s, commercialized in 2004 (Stirzaker, 2003; Stirzaker et al., 2010), and introduced by TISA into the study areas. It comprises a funnel-shaped device, two of which are buried at approximately one-third and two-thirds of the crop's root depth (Figure 8). As water percolates through the soil profile it collects in one or both funnels depending on the amount of water applied, soil type and initial soil moisture. The above ground indicator pops up when a funnel is filled with water. A rubber tube and syringe are used to extract the water, which is tested for

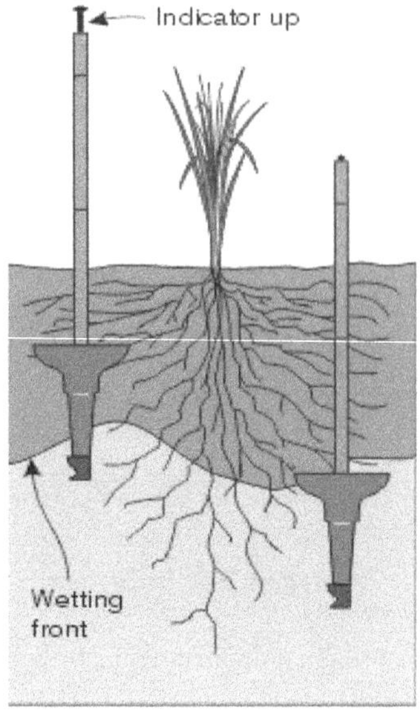

Figure 8. Placement of Wetting Front Detectors (VIA, 2019).

nitrates and salinity using colour test strips and an electrical conductivity metre. Hence, it illustrates to farmers the direct relationship between overwatering (consistent blue colours) and leaching of nutrients.

Processes of introducing and implementing the tools

In this project, the first step was to give TISA's project teams a thorough grounding in the use of the tools, which was done during a week-long workshop at project inception. Next, the recipients of the tools were selected. This process is very important as these farmers play a critical role in the spread of learning and development of social capital. The selection was based on discussions with local leaders and farmers using the following criteria; they must be respected and trusted within the scheme, distributed along the water delivery system (upstream, middle and downstream), and able to learn how to use the tools and communicate the learning to other farmers.

The scheme's extension officer (government farming advisor) and farmers were then trained in the installation and use of the tools and they installed the tools together. Between 2014 and 2015, Chameleon sensor arrays and Wetting Front Detectors were provided to 20 farmers, together with two Chameleon readers for each scheme. The depth of deployment varied depending on the crop (Table 4).

Table 4. Irrigation method, crops and placement depths of WFDs and Chameleon sensors (Stirzaker et al., 2017).

Scheme	Irrigation method	Crops monitored	Wetting front depths (cm)	Initial sensor depths (cm)
Kiwere	Gravity flood	Tomatoes	20, 50	20, 30, 40, 50
Silalatshani	Gravity flood	Maize	20, 40	15, 30, 45, 60
25 de Setembro (Boane)	Pump flood	Maize, cabbage	20, 40	15, 30, 45, 60

The farmers using the tools were provided with a field book and together with the extension officers were trained in recording the readings from the tools and data such as inputs used, quantity of crop harvested, prices of inputs and outputs, and other farm operations undertaken. The field books were critical in supporting farmer learning by allowing farmers to follow the changing wetting patterns, crop response to management decisions, and compute gross margins. The process of data collection varied: at i) Silalatshani people were hired to collect the data and help farmers fill in their field books; ii) Kiwere the scheme extension officer and scheme secretary collected the data; and iii) 25 de Setembro the IMC was responsible for the monitoring. Initially, the Chameleon readings were taken once or twice a week, with nitrate and salt recorded when the Wetting Front Detector indicator rose. The project's field staff manually entered the data from the field books into an Excel spreadsheet, which was sent to the country research team for local analysis and to the project research team for further analysis and cross-country comparisons.

The farmers were trained on how to use the tools only: i.e., what the Chameleon colours meant, when and how to extract water from the Wetting Front Detector, and how to analyse the water samples. This was a purposeful part of the research design as even though there may be a desirable colour pattern from a scientific perspective, this may differ from a farmer's perspective. Their decision-making is a function of their learning that is context dependent as they face a set of unique and complex constraints, including different appetites for risk (Bjornlund et al., 2019).

Group interaction can enhance the experiential learning from the tools; for example, focus groups were held with farmers at Kiwere to discuss how learning from the tools had changed farmers' behaviour; at 25 de Setembro, farmers initiated an AIP meeting to discuss their problems; and at Silalatshani informal weekly or bi-weekly discussions about the monitoring data took place between the project staff and farmers. These processes helped to consolidate the farmers' understanding of the water, fertilizer and soil dynamics and stimulated farmers to use the tools with different crops and at different depths to increase further their understanding of these dynamics and their trust in the tools.

Demonstration plots can also help to spread the knowledge and benefit of the tools beyond those directly using them. At Silalatshani, the AIP-initiated demonstration plots (Table 3) and farmers noted that during the first 3–4 weeks of maize crop growth, there was no need to irrigate up to the 40 cm zone as the roots were shallow. As a result, while only 24% of farmers surveyed in 2017 had the tools 73% changed their irrigation practices, and irrigators from other blocks started to show an interest in having the tools (Moyo et al., 2018).

At the time of this research, the Chameleon tool was in a pre-commercial development phase. It was important to have support processes in place – between users, inventors and producers of the tools – to resolve problems as they emerged. These included introducing the tools to the community to avoid the problems encountered in 25 de Setembro where 40% of the sensors were destroyed due to their cables being pulled from the fields because the community around the scheme was not involved in the installation. Also, in Kiwere they were pulled out to use as radio/TV antennas. There were also several technical problems with the tools due to the way farmers used them, such as loose connections, battery failure and failure to upload the data. User feedback enabled the inventor to improve these tools. At Silalatshani, the cost of employing people to take the readings, record data and update the database was problematic. In response, two farmers were trained to collect and record data on a weekly basis assisted by AGRITEX and DoIRR staff. This created an additional benefit by increasing the speed of farmer learning. Farmers also had problems correctly charging the solar-powered Chameleon readers; again, it was critical that technical backup was available. Land preparation disturbed the tools and protective structures were built around them in response. Discussions with farmers and further training ensured farmers took responsibility for the tools, which enabled the tools to last longer and provided opportunities to respond to farmers' questions related to understanding and interpreting the results. To ensure the continuation of the use of the tools, the project provided replacement tools when damage and malfunction occurred. Without such iterative development processes that engage users, SWM technologies may fail in developing countries.

In conclusion, it is important that there is clarity of responsibility for the tools, and farmers have back up in interpretation and problem-solving. However, the level of flexibility in the irrigation system is a determining factor as to whether farmers and the wider system can benefit from the tools and the learning that ensues. If farmers do not have the ability to manage their irrigation schedule based on the information from the tools, their value is limited. In this project, the information from tools and the AIP engagement of decision makers was critical in facilitating more flexible irrigation scheduling.

Synergies between the AIP and the tools: creating feedback loops

For farmer-level learning and decision-making, the tools and AIP approach worked well and behavioural changes occurred very quickly. By using the tools, farmers gained a deep understanding of the water-nutrient dynamics, which allowed them to make more informed decisions about water and nutrient management to avoid water stress, waterlogging and fertilizer leaching, and better utilize rainfall to reduce irrigation (Chilundo et al., 2020; Mdemu et al., 2020; Moyo et al., 2020; Stirzaker et al., 2017). These changes increased crop yields. In the context of adoption, this shows that the SWM technology has been adopted and the learning from the tools has enabled farmers to adapt their farming practices. Despite attempts to limit the cost of the technology, the willingness to pay literature pertaining to the tools suggests the need for some co-investment to ensure widespread adoption (Abebe et al., 2020; Adimassu et al., 2020).

As intended, the tools have triggered experimentation and are realizing important benefits for farmers, such as increased yields. A study on the use of the WFD in Ghana (where it was introduced without the Chameleon and without a supporting entity such as the AIP) indicated that a reasonable portion of farmers did not find the WFD easy to use (Adimassu et al., 2020). This suggests that deeper learning from the monitoring data requires additional processes. In the context of TISA, the AIP provided the initial platform for discussions and interpretation, and triggered increased farmer-to-farmer learning, experimentation, and adaptation of new irrigation regimes, which were quickly adopted and further increased yields (Parry et al., 2020). Group learning and demonstration plots initiated through the AIP have been important to augment the learning from the tools, thus overcoming the issues experienced in Ghana. As growing conditions and context are variable, the co-construction of knowledge is particularly valuable in the non-linear model of technology adoption.

Increased yield does not necessarily result in increased profitability or bring about systemic change. In this project, the AIPs facilitated the system's actors to address the additional constraints that prevent farmers from translating the improved yield into increased profits. For example, this was achieved at Silalatshani by providing gross-margin analysis training to extension officers who then realized the implications of enforcing a cropping calendar that stipulated subsistence crops with low profits. Based on this, the cropping calendar was removed and, with training in gross-margin analysis, farmers could make more informed crop choices and introduce higher-value crops and improved varieties. With improved links to markets, farmers could better match crop selection to market demand, engage in collective bargaining, and access better quality inputs. This combination was critical in translating increased yield into increased profitability (Chilundo et al., 2020; Mdemu et al., 2020; Moyo et al., 2020) and ensuring longevity of impact. As envisaged by TISA, this indicates that an AIP can assist in overcoming the socio-institutional and economic barriers that limit profitability.

The influence model emerging from the two-pronged approach shows some of the feedback loops in operation within these complex systems, the insights that farmers gained, and how they then initiated changes, which in turn provides new feedback in the system and additional changes and benefits (Figure 9). This model demonstrates the breadth and depth of change that has been experienced at the household and scheme scale. One and a half to three times as many farmers are aware of the tools as were provided with them. More than 50% of farmers have changed the frequency of their irrigation practice and the interval between irrigation events has increased.

Overall, households have experienced improved yields (an increase of 25% or more for 60% of households) and increased income (for 21% to 83% of households) (Bjornlund et al., 2018). These are significant outcomes. Increased yield and income were anticipated outcomes, but the feedback farmers received, in the form of reduced labour needed for irrigation, became a further driver of change. This allowed farmers to invest time in both farm and off-farm activities, further diversifying their income stream and increasing their resilience to shocks in the system.

Another critical feedback loop was created as upstream farmers reduced their irrigation frequency and/or duration, leading to a more reliable and adequate supply for downstream users at Kiwere (Tanzania). A flow-on effect is that downstream users' income has increased, conflict over access to water has reduced, and there is increased trust among irrigators. In response, farmers are now more willing to engage in collective action such as participation in irrigation maintenance and other infrastructure, and collective bargaining for inputs and outputs (Bjornlund et al., 2018). Increased household income has also reduced the level of conflicts between husbands and wives over how to prioritize the spending of scarce financial resources (Bjornlund et al., 2019).

In considering the conditions of the adoption model proffered by Perret and Stevens (2006), there are several key observations that can be made from the outcomes of the TISA project. In all schemes, the learnings from the process of introducing and using the SWM technology provided immediate incentives to change irrigation practice due to saved time and increased yield; hence, the tools brought about change in the farming system. The AIP augmented the learning from the tools and initiated solutions to other systemic barriers to

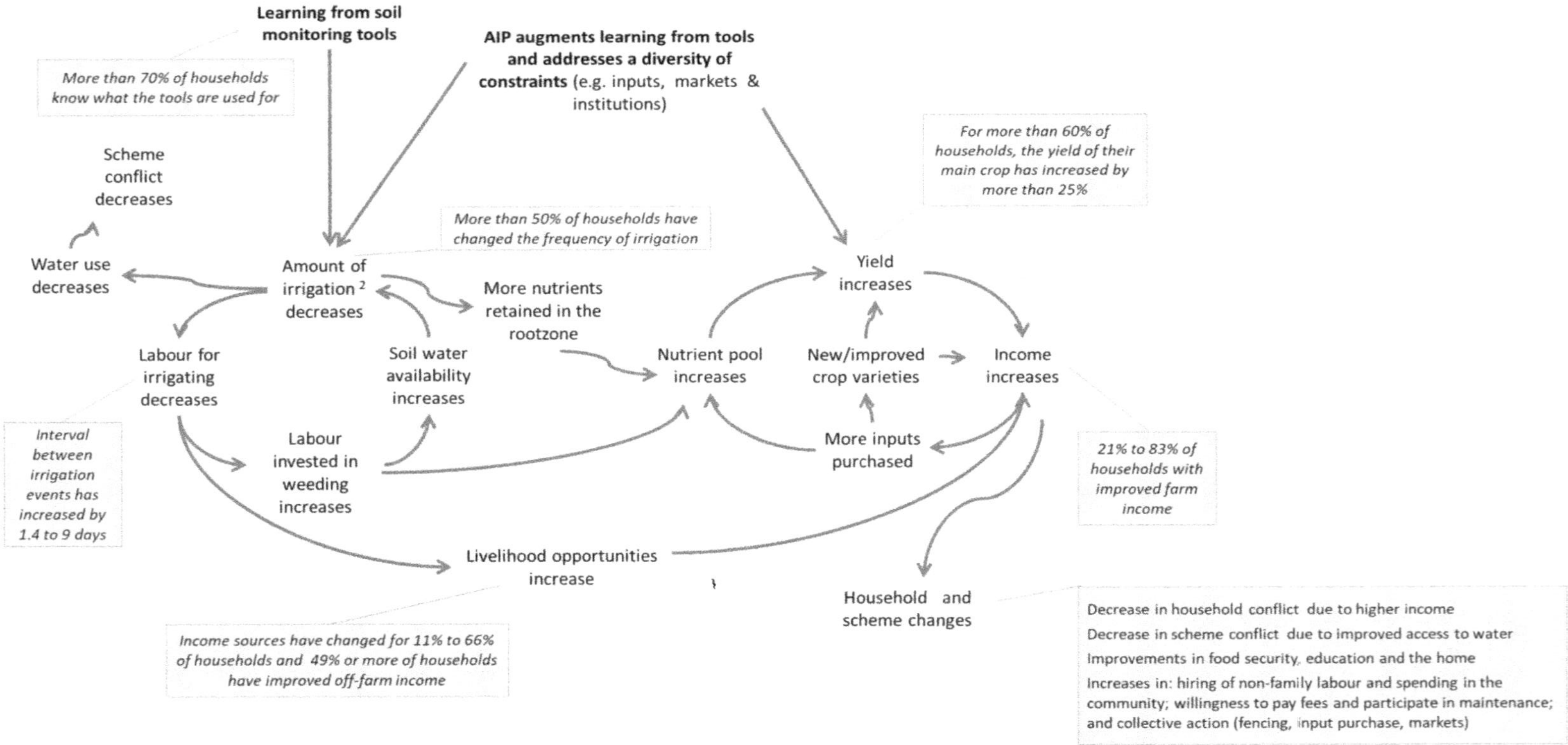

Figure 9. Influence model of monitoring tools and AIPs with key impacts[1] (adapted from Van Rooyen & Moyo, 2017a). [1]Outcomes from TISA's first phase (2014–2017) for Silalatshani, Kiwere and 25 de Setembro (Cheveia et al., 2018; Mdemu et al., 2018; Moyo et al., 2018); ranges of percentages are per scheme; [2]an aggregate term representing frequency, length of time, and/or volume.

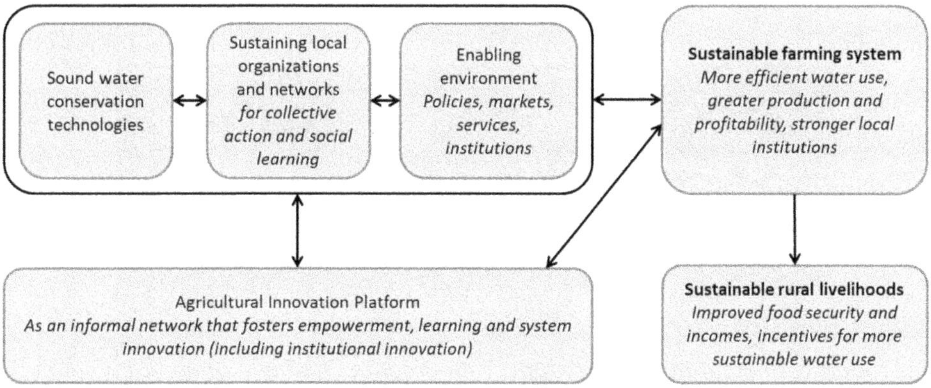

Figure 10. Conditions for water conservation technology adoption towards sustainable farming systems and rural livelihoods (Source: adapted from Perret & Stevens, 2006).

profitability, providing an important supplementary condition for adoption and ensuring that increased production translated into improved profitability. Social learning has spread the benefits of using the tools across the schemes, highlighting the importance of networking and social learning as an element of sustaining social conditions (Parry et al., 2020). Collective action was largely absent from the schemes at the project's inception, but water saving has reduced conflict and, combined with increased profitability of irrigation, has stimulated collective action and affirmed the relationship between technologies and collective action (Moyo et al., 2020). Perret and Stevens (2006) suggest that community-level collective action is important for the feasible introduction of technologies and these patterns should be investigated prior to implementation. However, adoption and change from the tools in the TISA project occurred quickly, so in this circumstance strong and prior collective action conditions appear not to have been a critical factor in adoption. As irrigation became more profitable through the AIP's activities, this stimulated further collective action, manifesting in actions such as a willingness to pay fees, self-organized maintenance and collective bargaining activities. This affirms that collective action is important for sustained adoption and transition to a more adaptive system.

We offer an amended model of adoption as experienced through the TISA project. Networks and social learning are added, as they are worthy of more explicit attention in the conditions for adoption. We add the AIP into the conditions of adoption due to its role in directly supporting and strengthening the value of the technology and collective action, and in initiating feedback and significant institutional innovation. The AIP could be described as the 'glue' between the elements in the adoption model. Acknowledging that AIPs may or may not need to persist in the long term, the AIP is kept as a separate additional enabling condition for technology adoption. As Figures 9 and 10 both confirm, the adoption process is complex and far removed from being a linear process.

Modifications to the tools and the process – making the tools 'smarter'

Data from the tools are potentially beneficial at two levels: for farmer learning and decision-making and for system learning for the benefit of scheme management,

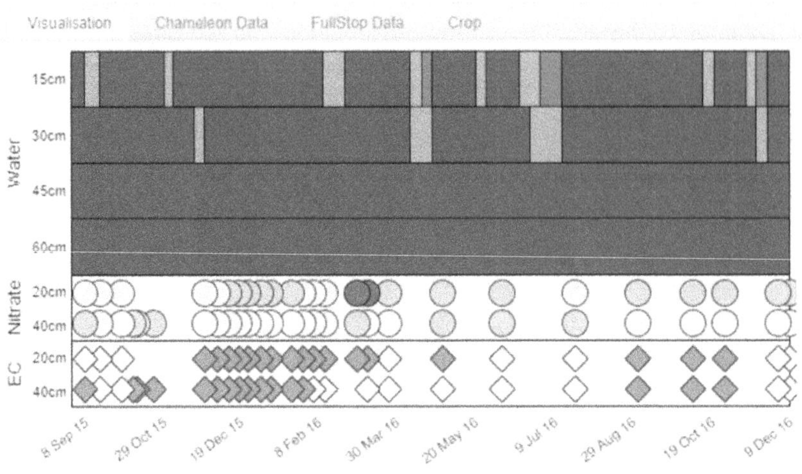

Figure 11. Seasonal patterns of soil moisture, nitrate and electrical conductivity for a sugar bean crop on Silalatshani (Source: VIA database, 2019). For nitrate, dark to lighter purple means high to low nitrate levels; EC measures dissolved salts: white indicates salinity of less than 0.4 (no salinity) green indicates 0.4 to 1.0 dS/m (low salinity); yellow 1.2–2.0 dS/m (low to medium salinity), orange 2.4–3.6 dS/m (medium to high salinity), red 4.0 to 5.2 dS/m (high salinity).

catchment management, and scientists. The manual monitoring and recording system worked well for the individual farmers as the critical information was available to them in real time in their fields to inform their decision-making. However, this was not satisfactory for the cross-scheme and country researchers due to: i) the slow pace at which data were transferred from the field, to country researchers and then to project researchers; ii) data accuracy, validity, transcription and quality; and iii) sensor locations becoming confused when farmers moved sensors.

Hence, a new Wi-Fi-enabled version of the reader was developed at CSIRO and introduced during year three of TISA and is currently being out-scaled as part of TISA and a second ACIAR funded project 'the Virtual Irrigation Academy'. This required some additional farmer training, but it reduced human error, enabled data problems to be fixed more quickly, and reduced the time for sharing data with the research team. This change did not affect the farmers' access to information and immediate decision-making. The new reader is paired with a smartphone and in addition to showing the colours, it also stores the data against each array's ID and transfers this to the smartphone via the hotspot function. When the smartphone has internet access, the data is uploaded to an online database, which is hosted by the Virtual Irrigation Academy (VIA platform) which became operational in 2016 (Virtual Irrigation Academy [VIA], 2019).[2] If there is no smartphone in the field the reader stores the data from multiple arrays and uploads it when Wi-Fi is available. The database can process and display the seasonal colour pattern for each farmer online as repeat measurements are recorded and uploaded (Figure 11).

The colours show the wetting and drying of the soil at four different depths (i.e., original Chameleon). The pattern shows that most root activity is in the 0–30 cm zone, but the soil profile is saturated below this depth for the whole season. At the beginning of the season, nitrate availability is below the root zone. The fertilizer added in December and March replenishes the profile but is lost from the 40 cm depth towards the end of the season. Salt levels at Silalatshani's Landela block are very low and do not constitute a problem.

Data related to inputs, yields and gross margins could also be manually uploaded to VIA, allowing analysis of soil moisture and nutrient trends within and across seasons, and how this influences farming outcomes. TISA still contributes to the accumulation of data from farmers and schemes around the world, which CSIRO are consolidating on the VIA platform. Thus, the VIA platform is creating a large database, which has the potential to be an important tool for water planners, authorities, irrigation managers, funding agency and policy makers and thereby create public benefits.

Out-scaling the smart water management technologies

The initial research (TISA 2013–17) at three government-owned smallholder irrigation schemes reported here had several limitations. In particular, the intensive resources devoted to interventions at the three irrigation schemes could not be sustained in business-as-usual irrigation farming. Consequently, following this prototype of a two-pronged SWM approach it is necessary to test if it can practically be out- and up-scaled to attain broader societal impact. This research is ongoing as part of TISA (2017–21).

The AIP process as part of TISA (2013–17) was resource intensive both in terms of financial and human capital. TISA (2017–21) is, therefore, testing whether the AIP processes can be undertaken in a more cost-effective manner to inform innovation at adjacent irrigation schemes (out-scaling) or at greater scales of governance (up-scaling) to assist more farmers. Such interventions at district and provincial government scales may permanently enhance existing governance arrangements rather than being time-limited AIP processes. This project used AIPs but other inclusive approaches may also provide the supportive processes needed to facilitate SWM technology to become farmer-owned, sustainable and profitable (for example, out-grower schemes, learning platforms). Opportunities, partnerships and administrative arrangements for up- and out-scaling will vary across countries.

Discussion and conclusion

The tools and AIP process triggered a series of systemic changes in the wider system. The tools provided information for farmers to improve the agronomic efficiency of the system, facilitated farmer learning and increased their confidence, and reduced conflict. The AIP enhanced the process of translating learning into changed practices and facilitated institutional innovation: for example, loosening adherence to the cropping calendar and growing more profitable crops; changing irrigation schedules; and developing linkages to input and output markets, which served as vital feedback mechanisms (Van Rooyen et al., 2020). The project thus confirmed the importance of addressing systemic or institutional barriers to technology adoption identified in the literature.

Initially, reductions in the frequency of irrigation were most probably based on the premise that reduced irrigation would maintain fertilizer within the root zone or at least not 'waste' fertilizer. As farmers began to 'skip' irrigation events, they gained confidence in the process (their crops survived) and soon skipping irrigation events translated into significant time savings (another critical incentive to change behaviour), which resulted in more time spent weeding and engaging in other livelihood opportunities (Chilundo et al., 2020; Mdemu et al., 2020; Moyo et al., 2020). At the end of the season, farmers realized that this process resulted in increased yields. Our findings provide support for the literature that stresses the importance of access to information both prior to and subsequent to adoption, and the need for a process of adoption and adaptation (Foster & Rosenzweig, 2010; Perret & Stevens, 2006; Wang et al., 2016). However, increased yield is unlikely to result in increased income without improved access to inputs and higher prices for crops; hence, the AIP linked farmers to input and output markets, creating the environment for increased incomes and providing a critical incentive for farmers to change behaviour. This process cemented the value of the information, which has resulted in real improvements in both the efficiency and the profitability of the irrigation scheme. Therefore, we conclude that the SWM tools provided critical information for learning and behavioural change that increased crop yields. Whereas the AIP increased profitability, which enhanced the incentive for behavioural change and the development of feedback loops that will maintain the new positive strategies. As highlighted in the literature, feedback in the system shapes actors' interactions, facilitates learning and increases adaptive capacity and resilience (Abson et al., 2017; Meadows, 2008). Increased profitability is critical for adoption supporting Michler et al.'s (2018) assertion that economic returns and profitability are important in addition to physical returns. TISA research shows a link between adoption and economic incentives, with benefits from the tools including reduced costs for pump fuel, fertilizer and labour and an emerging willingness to pay for the tools (Abebe et al., 2020; Chilundo et al., 2020; Mdemu et al., 2020; Moyo et al., 2020).

Moreover, the tools and AIP approach provided credibility for processes to change components of the system: for example, justifying farmers' negotiations of new and more flexible cropping and irrigation regimes with their irrigation management committee, which has ultimately resulted in real water savings. This creates yet another incentive for change as more water availability in night-storage dams (Silalatshani) and more water for downstream irrigation schemes (Kiwere) has reduced conflict and increased the willingness to participate in collective action. With respect to Perret and Stevens's (2006) model of SWM adoption, the findings emphasize the value of externally initiated multi-stakeholder processes, such as an AIP, to create or enhance the enabling conditions for SWM technology adoption and social learning.

These examples serve to illustrate the value of understanding the relationship between interventions in a complex system and the incentive framework of actors, which may lead to behavioural changes. It illustrates how the value of SMW interventions is strengthened in concert with institutional innovation processes. These processes should enable an environment where information can be used to increase system efficiencies. Most importantly, it improves the livelihood of smallholder farmers and their ability to feed their family more nutritional food and spend more money of the health and education of their children.

Notes

1.. Unless otherwise noted, this section draws predominantly from Van Rooyen et al. (2017b) and Bjornlund et al. (2018).

2.. VIA development is the focus of another ACIAR project '*A Virtual Irrigation Academy to improve Water Productivity in Malawi, Tanzania and South Africa*' managed by CSIRO Land and Water.

Disclosure statement

No potential conflict of interest was reported by the authors.

Funding

This study was part of the project 'Transforming small-scale irrigation in southern Africa' partially funded by the Australian Government via the Australian Centre for International Agricultural Research (FSC-2013-006 and LWR/2016/137). The project was also supported by the CGIAR Research Program on Water, Land and Ecosystems (WLE) and CGIAR Fund Donors.

ORCID

H. Bjornlund http://orcid.org/0000-0003-3341-5635
A. van Rooyen http://orcid.org/0000-0003-2035-049X
K. Parry http://orcid.org/0000-0003-0045-6967
M. Moyo http://orcid.org/0000-0002-5496-7554
M. Mdemu http://orcid.org/0000-0002-8110-8456
W. de Sousa http://orcid.org/0000-0002-7464-666X

References

Abdulai, A., & Huffman, W. (2014). The adoption and impact of soil and water conservation technology: An endogenous switching regression application. *Land Economics*, *90*(1), 26–43. https://doi.org/10.3368/le.90.1.26

Abebe, F., Wheeler, S., Zuo, A., Bjornlund, H., van Rooyen, A., Pittock, J., Mdemu, M., & Chilundo, M. (2020). Irrigators' willingness to pay for the access to soil moisture monitoring tools in South Eastern Africa. *International Journal of Water Resources Development*, 1–22. https://doi.org/10.1080/07900627.2020.1755956

Abson, D., Fischer, J., Leventon, J., Newig, J., Schomerus, T., Vilsmaier, U., von Wehrden, H., Abernathy, P., Ives, C., Jager, N., & Land, D. (2017). Leverage points for sustainability transformation. *Ambio*, *46*(1), 30–39. https://doi.org/10.1007/s13280-016-0800-y

Adimassu, Z., Balana, B. B., Appoh, R., & Nartey, E. (2020). The use of the wetting front detector as an irrigation-scheduling tool for pepper production in the Upper East Region of Ghana: Evidence from field experiment and farmers' perceptions. *Irrigation and Drainage*, 1–18. https://doi.org/10.1002/ird.2454

Akudugu, M. A., Guo, E., & Dadzie, S. K. (2012). Adoption of modern agricultural production technologies by farm households in Ghana: What factors influence their decisions? *Journal of Biology, Agriculture and Healthcare*, *2*(3), 1–14. https://www.iiste.org/Journals/index.php/JBAH/article/view/1522

Andersson, J., & d'Souza, S. (2014). From adoption claims to understanding farmers and contexts: A literature review of Conservation Agriculture (CA) adoption among smallholder farmers in southern Africa. *Agriculture, Ecosystems and Environment*, *187*, 116–132. https://doi.org/10.1016/j.agee.2013.08.008

Bjornlund, H., & Klein, K. K. (2015). Water conservation and trading: Policy challenges in Alberta Canada. In K. Burnett, R. Howitt, & J. Roumasset (Eds.), *Handbook of water economics and institutions* (pp. 381–396). Routledge.

Bjornlund, H., Pittock, J., Stirzaker, R., van Rooyen, A., Parry, K., Moyo, M., Zuo, A., Christie, E., Mdemu, M., de Sousa, W., Cheveia, E., Munguambe, P., Kimaro, E., Kissoly, L., & Chilundo, M. (2018). Transforming smallholder irrigation into profitable and self-sustaining systems in southern Africa. In IWRA and K-Water (Ed.), *Smart water management: Case study report* (pp. 239–385). Korean Water Resources Corporation (K-water) and the International Water Resources Association (IWRA). https://www.iwra.org/wp-content/uploads/2018/11/SWM-report-final.pdf

Bjornlund, H., van Rooyen, A., & Stirzaker, R. (2017). Profitability and productivity barriers and opportunities in small-scale irrigation schemes. *International Journal of Water Resources Development*, *33*(5), 690–704. https://doi.org/10.1080/07900627.2017.1326881

Bjornlund, H., Zuo, A., Wheeler, S., Parry, K., Pittock, J., Mdemu, M., & Moyo, M. (2019). The dynamics of the relationship between household decision-making and farm household income in small-scale irrigation schemes in southern Africa. *Agricultural Water Management*, *213*, 135–145. https://doi.org/10.1016/j.agwat.2018.10.002

Cai, X., Molden, D., Mainuddin, M., Sharma, B., Ahmad, M.-D., & Karimi, P. (2011). Producing more food with less water in a changing world: Assessment of water productivity in 10 major river basins. *Water International*, *36*(1), 42–62. https://doi.org/10.1080/02508060.2011.542403

Cheveia, E., de Sousa, W., Faduco, J., Mondlhane, E., Chilundo, M., Tafula, M., & Christen, E. (2018). *Adoption and impacts of irrigation management tools and agricultural innovation platforms (AIP) in Mozambique: Mozambique report on the final survey of the project* (FSC-2013-006). Australian National University.

Chilundo, M., de Sousa, W., Christen, E. W., Faduco, J., Bjornlund, H., Cheveia, E., Munguambe, P., Jorge, F., Stirzaker, R., & van Rooyen, A. F. (2020). Do agricultural innovation platforms and soil moisture and nutrient monitoring tools improve the production and livelihood of smallholder irrigators in Mozambique? *International Journal of Water Resources Development*, 1–21. https://doi.org/10.1080/07900627.2020.1760799

Choi, G. W., Chong, K. Y., Kim, S. J., & Ryu, T. S. (2016). SWMI: New paradigm of water resources management for SDGs. *Smart Water*, *1*(1), 1–12. https://doi.org/10.1186/s40713-016-0002-6

Corbeels, M., de Graaf, J., Ndah, T. H., Penot, E., Baudron, F., Naudin, K., Andrieu, N., Chirat, G., Schuler, J., Nyagumbo, I., Rusinamhodzi, L., Traore, K., Mzoba, H. D., & Adolwa, I. S. (2014). Understanding the impact and adoption of conservation agriculture in Africa: A multi-scale analysis. *Agriculture, Ecosystems and Environment*, *187*, 155–170. https://doi.org/10.1016/j.agee.2013.10.011

De Sousa, W., Ducrot, R., Munguambe, P., Bjornlund, H., Machava, A., Cheveia, E., & Faduco, J. (2017). Irrigation and crop diversification in the de Setembro irrigation scheme, Mozambique. *International Journal of Water Resources Development*, *33*(5), 705–724. https://doi.org/10.1080/07900627.2016.1262246

Fanadzo, M., Chiduza, C., & Mnkeni, P. (2010). Overview of smallholder irrigation schemes in South Africa: Relationship between farmer crop management practices and performance. *African Journal of Agricultural Research*, *5*(25), 3514–3523. http://www.academicjournals.org/AJAR

Foster, A. D., & Rosenzweig, M. R. (2010). Microeconomics of technology adoption. *Annual Review of Economics*, *2*(1), 395–424. https://doi.org/10.1146/annurev.economics.102308.124433

Garb, Y., & Friedlander, L. (2014). From transfer to translation: Using systemic understandings of technology to understand drip irrigation uptake. *Agricultural Systems*, *128*, 13–24. https://doi.org/10.1016/j.agsy.2014.04.003

Glover, D., Sumberg, J., Ton, G., Andersson, J., & Badstue, L. (2019). Rethinking technological change in smallholder agriculture. *Outlook for Agriculture*, *48*(3), 169–180. https://doi.org/10.1177/0030727019864978

Haensch, J., Wheeler, S., & Zuo, A. (2019). Do neighbors influence irrigators' permanent water selling decisions in Australia? *Journal of Hydrology*, *572*, 732–744. https://doi.org/10.1016/j.jhydrol.2019.03.023

Independent Science and Partnership Council (ISPC). (2015). *Strategic study of good practice in AR4D partnership*. CGIAR Independent Science and Partnership Council (ISPC). https://cas.cgiar.org/sites/default/files/ISPC_StrategicStudy_Partnerships.pdf

Kim, H. (2019). *What is smart water management: An explainer*. Development Asia. https://development.asia/explainer/what-smart-water-management

Knickel, K., Brunori, G., Rand, S., & Proost, J. (2008, 6-10 July). *Towards a better conceptual framework for innovation processes in agriculture and rural development: From linear models to systemic approaches* [Paper presented]. European IFSA Symposium, Clermont-Ferrand, France.

Makini, F. W., Kamau, G. M., Makelo, M. N., & Mburathi, G. K. (2013). *A guide for developing and managing agricultural innovation platforms*. Kenya Agricultural Research Institute and Australian Centre for International Agricultural Research.

Maru, Y. (2018). Editorial Introduction: Critical reflection on and learning from Agricultural Innovation Systems (AIS) approaches and emerging Agricultural Research for Development (AR4D) practice. *Agricultural Systems*, *165*, 294–295. https://doi.org/10.1016/j.agsy.2018.07.011

Mdemu, M., Kissoly, L., Bjornlund, H., Kimaro, E., Christen, E., Stirzaker, R., Stirzaker, R., & Ramshaw, P. (2020). The role of soil monitoring tools and agricultural innovation platforms in improving food security and income of farmers in smallholder irrigation schemes in Tanzania. *International Journal of Water Resources Development*, 1–23. https://doi.org/10.1080/07900627.2020.1765746

Mdemu, M., Kissoly, L., Kimaro, E., Bjornlund, H., & Parry, K. (2018). *Increasing irrigation productivity in Mozambique, Zimbabwe and Tanzania through on farm monitoring, adaptive management and agriculture innovation platforms: End-of-project survey Tanzania* (FSC-2013-006). Australian National University.

Mdemu, M., Mziray, N., Bjornlund, H., & Kashaigili, J. (2017). Barriers to and opportunities for improving productivity and profitability of the Kiwere and Magozi irrigation schemes in Tanzania. *International Journal of Water Resources Development*, *33*(5), 725–739. https://doi.org/10.1080/07900627.2016.1188267

Meadows, D. (2008). *Thinking in systems: A primer*. Earthscan.

Michler, J. D., Tjernstrom, E., Verkaart, S., & Mausch, K. (2018). Money matters: The role of yields and profits in agricultural technology adoption. *American Journal of Agricultural Economics*, *101*(3), 710–731. https://doi.org/10.1093/ajae/aay050

Moyo, M., Maya, M., van Rooyen, A., Dube, T., Parry, K., & Bjornlund, H. (2018). *Increasing irrigation water productivity in Mozambique, Tanzania and Zimbabwe through on-farm monitoring, adaptive management and agricultural innovation platforms: End of project survey report for Zimbabwe*. Australian National University. https://africawaterproject.wordpress.com/updates/publications/

Moyo, M., van Rooyen, A., Bjornlund, H., Parry, K., Stirzaker, R., Dube, T., & Maya, M. (2020). The dynamics between irrigation frequency and soil nutrient management: Transitioning small scale irrigation towards more profitable and sustainable systems in Zimbabwe. *International Journal of Water Resources Development*, 1–25. https://doi.org/10.1080/07900627.2020.1739513

Moyo, M., van Rooyen, A., Moyo, M., Chivenge, P., & Bjornlund, H. (2017). Irrigation development in Zimbabwe: Understanding productivity barriers and opportunities at Mkoba and Silalatshani irrigation schemes. *International Journal of Water Resources Development*, *33*(5), 740–754. https://doi.org/10.1080/07900627.2016.1175339

Mutiro, J., & Lautze, J. (2015). Irrigation in Southern Africa: Success or failure? *Irrigation and Drainage*, *64*(2), 180–192. https://doi.org/10.1002/ird.1892

Mwangi, M., & Kariuki, S. (2015). Factors determining adoption of new agricultural technology by smallholder farmers in developing countries. *Journal of Economics and Sustainable Development*, *5*(5), 208–216. https://www.researchgate.net/publication/303073456

Myeni, L., Moeletsi, M., & Clulow, A. (2019). Present status of soil moisture estimation over the African continent. *Journal of Hydrology: Regional Studies*, *21*, 14–24. https://doi.org/10.1016/j.ejrh.2018.11.004

North, D. C. (1991). Institutions. *Journal of Economic Perspectives, 5*(1), 97–112. https://doi.org/10.1257/jep.5.1.97

Ostrom, E. (2005). Doing institutional analysis: Digging deeper than markets and hierarchies. In C. Menard & M. M. Shirley (Eds.), *Handbook of new institutional economics* (pp. 819–848). Springer.

Ostrom, E. (2007). A diagnostic approach for going beyond panaceas. *Proceedings of the National Academy of Sciences, 104*(39), 15181–15187. https://doi.org/10.1073/pnas.0702288104

Pamuk, H., Bulte, E., & Adekunle, A. A. (2014). Do decentralized innovation systems promote agricultural technology adoption? Experimental evidence from Africa. *Food Policy, 44*, 227–236. https://doi.org/10.1016/j.foodpol.2013.09.015

Park, J. Y., Hwang, G. S., Ryu, M., Mun, K. H., Yi, S., Lee, K., Kuisma, S., Nickum, J., Clench, C., Bjornlund, H., Stephan, R. M., & Bond, H. (Eds.). (2018). *Smart water management: Case study report*. Korean Water Resources Corporation (K-water) and the International Water Resources Association (IWRA). https://www.iwra.org/wp-content/uploads/2018/11/SWM-report-final.pdf

Parry, K., van Rooyen, A., Bjornlund, H., Pittock, J., Kissoly, L., Moyo, M., & Chilundo, M. (2020). The importance of learning processes in transforming small-scale irrigation schemes. *International Journal of Water Resources Development*, 1–25. https://doi.org/10.1080/07900627.2020.1767542

Perret, S. R., & Stevens, J. B. (2006). Socio-economic reasons for the low adoption of water conservation technologies by smallholder farmers in southern Africa: A review of the literature. *Development Southern Africa, 23*(4), 461–476. https://doi.org/10.1080/03768350600927193

Pittock, J., Bjornlund, H., & van Rooyen, A. F. (2020). Transforming failing small holder irrigation schemes in Africa: A theory of change. *International Journal of Water Resources Development*. Forthcoming.

Pittock, J., & Stirzaker, R. (2014). *Project proposal: Increasing irrigation water productivity in Mozambique, Tanzania and Zimbabwe through on-farm monitoring, adaptive management and agricultural innovation platforms* (FSC-2013-006). Australian National University.

Ramos, H. M., McNabola, A., López-Jiménez, P. A., & Pérez-Sánchez, M. (2020). Smart water management towards future water sustainable networks. *Water, 12*(58), 1–13. https://doi.org/10.3390/w12010058

Schut, M., Andersson, J., Dror, I., Kamanda, J., Sartas, M., Mur, R., Kassam, S., Brouwer, H., Stoian, D., Devaux, A., Velasco, C., Gramzow, A., Dubois, T., Flor, R., Gummert, M., Buizer, D., McDougall, C., Davis, K., Homann-Kee Tui, S., & Lundy, M. (2017). *Guidelines for innovation platforms in agricultural research for development: Decision support for research, development and funding agencies on how to design, budget and implement impactful innovation platforms*. International Institute of Tropical Agriculture (IITA) and Wageningen University (WUR) under the CGIAR Research Program on Roots Tubers and Bananas (RTB). https://edepot.wur.nl/420696

Stevens, J. B. (2006). *Adoption of irrigation scheduling methods in South Africa* [Doctoral thesis]. University of Pretoria.

Stevens, J. B., Duvel, G. H., Steyn, G. J., & Marobane, W. (2005). *The range, distribution and implementation of irrigation scheduling models and methods in South Africa* (WRC Report No. 1137/1/05). Water Research Commission.

Stirzaker, R., & Pittock, J. (2014). Chapter 5: The case for a new irrigation research agenda for sub-Saharan Africa. In J. Pittock, R. Q. Grafton, & C. White (Eds.), *Water, food and agricultural sustainability in southern Africa* (1st ed., pp. 91–104). Tilde Publishing and Distribution.

Stirzaker, R., Mbakwe, I., & Mziray, N. (2017). A soil and water solute learning system for small-scale irrigators in Africa. *International Journal of Water Resources Development, 33*(5), 788–803. https://doi.org/10.1080/07900627.2017.1320981

Stirzaker, R., Stevens, J., Annandale, J. G., & Steyn, J. M. (2010). Stages in the adoption of a wetting front detector. *Irrigation and Drainage, 59*(4), 367–376. https://doi.org/10.1002/ird.472

Stirzaker, R. J. (2003). When to turn the water off: Scheduling micro-irrigation with a wetting front detector. *Irrigation Science, 22*(3–4), 177–185. https://doi.org/10.1007/s00271-003-0083-5

Stirzaker, R. J. (2006). *Soil water monitoring: State of play and barriers to adoption* (Irrigation Matters Series 01/06). CRC for Irrigation Futures.

Sullivan, A., & Pittock, J. (2014). Chapter 3: Agricultural policies and irrigation in Africa. In J. Pittock, R. Q. Grafton, & C. White (Eds.), *Water, food and agricultural sustainability in southern Africa* (1st ed., pp. 30–54). Prahan: Tilde Publishing and Distribution.

Van Rooyen, A., & Moyo, M. (2017a, 29 May-2 June 2017). *The transition of dysfunctional irrigation schemes towards complex adaptive systems: The role of agricultural innovation platforms* [Paper presented], World Water Congress, Cancun, Mexico.

Van Rooyen, A., Moyo, M., Bjornlund, H., Dube, T., Parry, K., & Stirzaker, R. (2020). Identifying leverage points to transition dysfunctional irrigation schemes towards complex adaptive systems. *International Journal of Water Resources Development*, 1–28. https://doi.org/10.1080/07900627.2020.1747409t

Van Rooyen, A., Ramshaw, P., Moyo, M., Stirzaker, R., & Bjornlund, H. (2017b). Theory and application of agricultural innovation platforms for improved irrigation scheme management in southern Africa. *International Journal of Water Resources Development*, *33*(5), 725–739. https://doi.org/10.1080/07900627.2017.1321530

Van Rooyen, A., Swaans, K., Cullen, B., Lema, Z., & Mundy, P. (2013). *Facilitating innovation platforms* (Innovation Platforms Practice Brief 10). ILRI.

Virtual Irrigation Academy (VIA). (2019). *The virtual irrigation academy*. CSIRO Agriculture and Food. https://via.farm/

Wada, Y., Flörke, M., Hanasaki, N., Eisner, S., Fischer, G., Tramberend, S., Satoh, Y., van Vliet, M. T. H., Yillia, P., Ringler, C., Burek, P., & Wiberg, D. (2016). Modelling global water use for the 21st century: The Water Futures and Solutions (WFaS) initiative and its approaches, *Geoscientific Model Development*, *9*(1), 175–222. Copernicus Publications. https://www.geosci-model-dev.net/9/175/2016/

Wang, J., Bjornlund, H., Klein, K. K., Zhang, L., & Zhang, W. (2016). Factors that influence the rate and intensity of adoption of improved irrigation technologies in Alberta, Canada. *World Economics and Policy*, *2*(3), 1650026 1–32. https://doi.org/10.1142/S2382624X16500260

Wang, J., Klein, K. K., Bjornlund, H., Zhang, L., & Zhang, W. (2015). Adoption of improved irrigation scheduling methods in Alberta: An empirical analysis. *Canadian Water Resources Journal*, *40*(1), 47–61. https://doi.org/10.1080/07011784.2014.975748

Wheeler, S. A., Zuo, A., Bjornlund, H., Mdemu, M. V., van Rooyen, A., & Munguambe, P. (2017). An overview of extension use in irrigated agriculture and case studies in south-eastern Africa. *International Journal of Water Resources Development*, *33*(5), 755–769. https://doi.org/10.1080/07900627.2016.1225570

Woltering, L., Fehlenberg, K., Gerard, B., Ubels, J., & Cooley, L. (2019). Scaling – From "reaching many" to sustainable systems change at scale: A critical shift in mindset. *Agricultural Systems*, *176*, 102652. https://doi.org/10.1016/j.agsy.2019.102652

Using innovative smart water management technologies to monitor water provision to refugees

Ryan W. Schweitzer, Ben Harvey and Murray Burt

Introduction

Water supply and refugees

Globally, approximately 785 million people do not have access to a safe and reliable water source (UNICEF/WHO, 2019). Over twice that number, 1.9 billion, do not have access to water on their premises. The health impacts of drinking water from unimproved sources have been demonstrated (Clasen et al., 2014; GBD, 2016). However, there are also economic and human development impacts. The burden of water collection often falls on women and girls, with the time dedicated to water collection resulting in lower educational attainment and lower productivity (Koolwal & Van de Walle, 2013) and, critically, exposing them to a range of health hazards, including exposure to sexual, psychological, physical and sociocultural violence (Sommer et al., 2015).

Sustainable Development Goal 6 aims to 'ensure availability and sustainable management of water and sanitation for all' by the year 2030. One critical demographic that is at risk of being left behind are the 70 million people forcibly displaced from their homes as a result of conflict, persecution or human rights violations. Nearly one-third of these, 25 million, are refugees who have crossed an international border to seek protection (UNHCR, 2019). For them, as for the rest of humanity, access to safe drinking water is fundamental for survival and a basic human right guaranteed by international law. In 2010 the UN General Assembly, through Resolution 64/292, explicitly recognized the human right to water and sanitation, building on previous international treaties. About 71% of the global population has access to safely managed drinking water services on premises (UNHCR, 2020a). For refugees, the figure is half that, only 35% at the end of 2018. The remaining 65% of refugees, approximately 12.9 million people, do not have access to clean tap water.

The United Nations High Commissioner for Refugees

The UN High Commissioner for Refugees (UNHCR) is the agency mandated with ensuring the international protection of refugees, asylum seekers, internally displaced and stateless persons. In 2019 UNHCR had a budget of USD143.8 million to support the delivery of water supply services in programmes across 42 countries, with most of these expenditures focused on the over 165 camps and settlements that UNHCR together with its partners manages (UNHCR, 2020b).

UNHCR and its partners face a range of challenges to ensure the right to water is realized for all persons of concern as well as the host community members. Challenges can be structural (e.g. increasing numbers of displaced persons, insufficient financial and human resources), environmental (e.g. water scarcity and the effects of climate change), or operational (e.g. lack of supply chains, security issues for accessing remote areas). Faced with these challenges, UNHCR and its partners have been exploring how smart water monitoring technology can be leveraged to improve cost efficiency and programming effectiveness, and ultimately ensure improved health and dignity for displaced populations and the communities that host them. Smart water monitoring technology is more relevant than ever given the particular constraints caused by the coronavirus which have resulted in movement restrictions and site closures that pose a challenge to ensuring the continuity of water supply services in refugee camps and settlements.

Background to the Uganda case

Beginning in June 2016, many South Sudanese refugees crossed the border seeking protection in the West Nile region of northern Uganda. By January 2020, there were 867,453 registered South Sudanese refugees in Uganda (UNHCR, 2020c). At the beginning of the influx there were hardly any basic services, such as electricity, water or sanitation, outside of the small towns in this region of Uganda (UNHCR, 2017). The humanitarian response included the transport of potable water via tanker trucks (Figure 1) to ensure that the minimum threshold of 15 litres per person per day could be provided. At its peak, this required delivery of 6387 m^3 of water per day by 630 water tanker trucks (UNHCR, 2018).

The financial costs and logistical challenges of managing and monitoring the water delivery operation exceeded the capacity of UNHCR and its partners. An audit by the UN Office of Internal Oversight (OIOS, 2018) found severe risk management and control deficiencies. One key recommendation of the audit was that UNHCR develop a better system

Figure 1. A water tanker truck fills up a reservoir while refugees wait to collect water in Bidibidi Refugee Settlement in Northern Uganda.

for monitoring water delivery during water trucking operations to address financial risk management concerns, to ensure that UNCHR was paying for the actual services delivered, and above all that refugees' right to water was met and UNHCR could be accountable to the affected population and its donors. The management response to the audit was to investigate the application of smart water monitoring technologies in refugee emergencies.

Methodology

In 2018 UNHCR conducted a desk review of available water level monitoring technologies. This review identified six different technologies that could be used for smart water management (SWM): electro-resistive, float-based resistivity, piezometric, ultrasonic, radar, and wave-radar probe. Eight water monitoring devices from six different companies (TankMatix, HummBox, Libelium, Tekelek, DecentLabs and KFA) were field tested. These devices used data communication technologies that included GSM cellular technology (3G and 4G) as well as low power wide area network, specifically Long Range Wide Area Network (LoRaWAN), which is based on chirp spread spectrum modulation (LoRa Alliance, 2015). Previous studies have showed the advantages of LoRaWAN over other 'internet of things' technologies (Mekki et al., 2017). Table 1 has a list of the devices that were tested in the pilot.

Pilot testing in Uganda

A pilot field test was planned to further investigate the potential for SWM technologies in refugee situations. This test considered two different communication technologies and the following parameters of various water level sensors: accuracy and precision, device cost, reliability, and battery life. Devices were selected that could be installed in static reservoirs and were designed to link water level measurements to geographical location. In January 2019 field testing was carried out in Rhino Camp Refugee Settlement.

Description of field tests

The field accuracy of the water level devices was established by comparing a series of readings from the installed devices and comparing them against measurements from a manual well dip metre (Butyl BDM5/1-50 m).

Each of the monitoring devices needed to be configured before it could be deployed. For the GSM devices this involved installation of local SIM cards, setting the device's

Table 1. Devices included in the field test.

Manufacturer	Name (model number)	Communication Technology
Hummbox	Ultrasonic Sensor	LoRaWAN
Libelium	Plug and Sense	3G/4G
TankMatix	Fuel Tanker 3 G + GPS	3G
DecentLabs	Ultrasonic Sensor (DL-MBX-001)	LoRaWAN
DecentLabs	Piezometric Pressure Sensor (DL-PR36)	LoRaWAN
Tekelek	Ultrasonic Sensor (Tek766)	LoRaWAN
Tekelek	Ultrasonic Sensor (TEK733)	3G
KFA	Remote Level Sensor (KFA2)	LoRaWAN

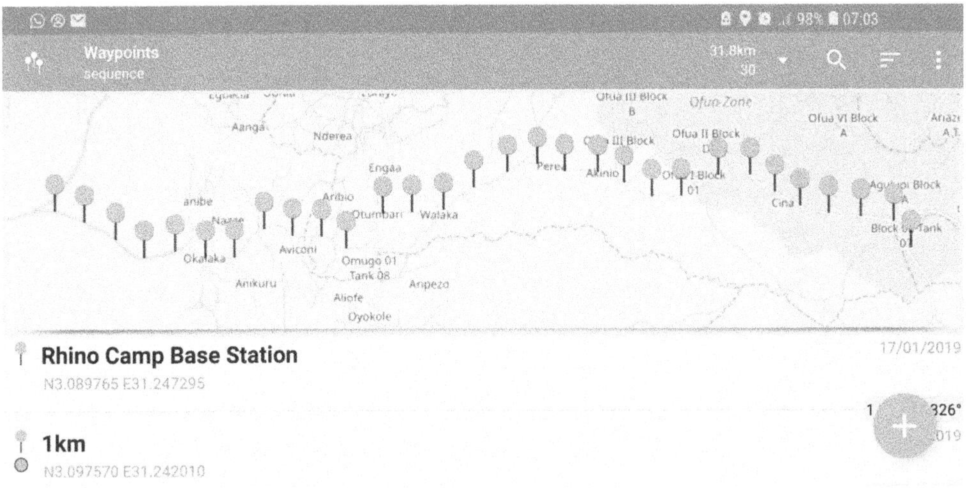

Figure 2. Locations used to verify LoRaWAN transmission range (pins every 1 km – max. range 29 km).

mobile internet access point name, and registering the GSM modem with the proprietary dashboard service. For the LoRaWAN-enabled devices this involved waking up the devices for first use and registering them with the LoRaWAN packet service (Things Network). All devices needed to be adjusted in the following way: changing the measurement frequency (from every 30 minutes to every 3 hours), changing the timing of measurements (selecting hours of operation, e.g. from 8 AM to 8 PM), and calibrating to the size of the reservoir within which the device was mounted. The process of configuration was documented and qualitatively rated by the UNHCR team on a scale from 1 (easiest) to 10 (hardest).

For the LoRaWAN-enabled devices, a range test was conducted by using a GPS tracker (Android GPS Utilities App) to measure the straight-line distance between the UNHCR vehicle and the base antenna where the LoRa gateway was installed. The UNHCR vehicle was stopped every kilometre away from the base antenna (line-of-sight distance, not road distance), and each of the various LoRaWAN devices (nodes) was tested to ensure that it was able to send data back to the gateway. Where 3G coverage was available, the water, sanitation and hygiene (WASH) consultant was able to verify that LoRaWAN data were being sent by checking the various device internet dashboards online, in real time. Figure 2 shows the locations where each of the devices was tested.

Results and discussion

Overall, the SWM devices piloted were accurate enough to allow the collection of meaningful real-time water level data – to monitor water provision and establish the basis for validation of payment to water trucking contractors. As an example of the operational utility of these SWM systems, Figure 3 shows the data collected from a 10,000-litre water reservoir from 3rd to 9th February 2020. It is clear that no water was delivered on 4th or 8th February.

The following section discusses the specific results of the testing done on the communication technology and water level sensors included in the pilot.

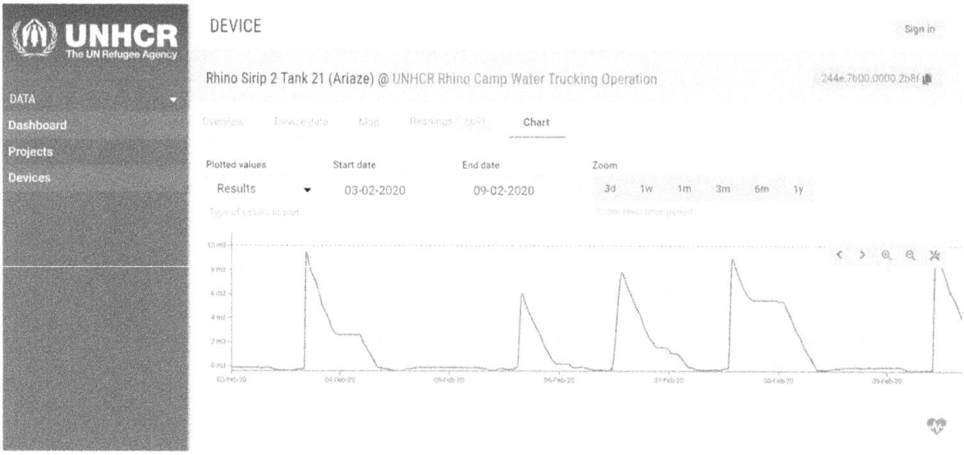

Figure 3. Sensor readings from from Rhino Camp using Tek766 LoRaWAN ultrasonic water sensor displayed on UNHCR's Dashboard.

Communication technologies

The pilot testing found that cellular communication technology is better suited for applications that require broader geographic coverage; however, these applications consume considerable power in measurement and require high data throughput, while LoRa is better suited for sensors that collect small amounts of data (up to 243 bytes per message) in areas without 3G or 4G coverage. It is also well suited for applications where individual measurements can be grouped into data 'packets' and transmitted less frequently, thus extending battery life, as demonstrated by Mekki et al. (2017). There are considerable economies of scale to utilizing LoRa-enabled sensors, since each gateway can connect to thousands of devices and handle over a million messages per day. Gateways that are compatible with the Things Network and have open source packet forwarding software allow the data to be pushed to the cloud, where it can be visualized from anywhere in the world with an internet connection. During the Uganda pilot, a single LoRaWAN gateway installed was able to collect data from LoRa water monitoring devices within an area of 1200 km^2 at ranges of up to 29 km.

The pilot also found that GSM devices worked well. However, based on the manufacturers' suggested retail prices for the respective devices, if there are more than three water reservoirs within a 1000 km^2 area it becomes significantly more cost-effective to opt for LoRaWAN installations. Economies of scale are achieved because a single gateway which costs between USD1500 and USD2000 can be mounted on a telecommunications tower and connected to the internet via ethernet cable, thereby supporting thousands of individual devices to spread the cost of the gateway, significantly lowering operational costs (no SIM cards and network fees), and enhancing the benefits of the LoRaWAN network to other sectors and partners.

Water level sensors

Both ultrasonic and piezometric sensors proved to be accurate (± 0.2–0.8 cm) in comparison to manual dip meter measurement. The ultrasonic devices are preferred, as they are a lower-cost option and do not need to be calibrated to barometric pressure. The sensor using radar was also accurate for static water level monitoring, but did not function well for the mobile water truck applications. The pulse radar encountered several equipment failures and overall did not perform as well as the other types of sensors.

Mobile versus static monitoring

Table 2 summarizes the performance of the water level sensors tested in this pilot. Water level monitoring in static reservoirs worked better than in mobile water tanker trucks, where the antennae of the device were externally exposed and susceptible to being damaged as the trucks brushed against trees and the like. In addition, these devices gave a considerable number of false readings, which were likely due to waves or ripples in the water as the tank moved.

Conclusion

The smart water monitoring technologies piloted in Uganda have considerable potential to revolutionize the delivery of water trucking services in emergency refugee responses. The technologies proved to be accurate, reliable and cost-effective. They can not only provide an important tool to reduce financial risk to UNHCR and other humanitarian actors, but also be a critical tool for ensuring that the refugees' right to water is maintained. Nevertheless, the Sustainable Development Goals target for water supply goes beyond quantity of water and requires ensuring that water is free from priority chemical and microbial contaminants and that the water is available on demand. This would require that additional parameters are monitored by the SWM system. For example, SWM systems that are designed to measure the flow rate in distribution systems to ensure that water is available on demand can also be designed to monitor water quality parameters such as free residual chlorine or turbidity at various points in the system. They can also be used as an asset management tool by operators to optimize the performance of water supply systems. For example, operators can compare the water table drawdown for various pumping regimes against the generation of solar. Water operators can monitor performance over time to identify seasonal demands or cumulative impacts on groundwater levels or pump performance to identify early symptoms of borehole ageing.

Future work

UNHCR has found SWM, in particular those technologies that utilize LoRaWAN for communication, to be easy, cheap, reliable and scalable for rapid monitoring of WASH services in refugee contexts. In June 2019 UNHCR began to work with partners to explore alternative applications of the SWM technologies. Several different types of

Table 2. Overview of the characteristics of the water level sensors included in the field test.

Water monitoring device	Hummbox	Libelium Plug and Sense	DecentLabs Ultrasonic	DecentLabs Pressure	Tekelek TEK766	Tekelek TEK733	KFA Remote Level Sensor
Technology	Ultrasonic	Radio	Ultrasonic	Piezometric	Ultrasonic	Ultrasonic	Electromagnetic
Transmission range	29 km	27 km	26 km	26 km	28 km	28 km	28 km
Accuracy	±0.2 cm	±0.2 cm	±0.5 cm	±0.3 cm	±0.3 cm	±0.8 cm	±0.2 cm
Battery life	Excellent	Solar power	Excellent	Excellent	Excellent	Good	Good
LoRaWAN enabled	Yes	Yes	Yes	Yes	Yes	No	No
Ease of configuration	8	3	9	9	5	3	4
Estimated unit cost[a]	$370	$970	$600	$790	$60	$100	$1,100

[a]May not reflect actual retail prices.

Table 3. Sensors to be included in further pilot testing.

Sensor	Application
Pressure probe	Monitor groundwater levels and impacts on the aquifer Monitor dynamic pumping conditions to protect electro-mechanical equipment and avoid damage to borehole (e.g., fouling of screen due to excessive drawdown) Measure pressure in distribution systems to detect leaks
Electricity meters	Measure performance of solar water pumping systems for system optimization or detect need for maintenance
Water quality	Monitor water quality parameters (free residual chlorine, turbidity, pH, electro-conductivity, temperature, etc.) to make required adjustments to water treatment systems or flag
Flow meters	Monitor consumption to ensure basic needs are met

sensors were identified to be piloted. Table 3 lists these sensors as well as the associated application for each.

By the end of 2020, UNHCR aims to instal over 1300 SWM devices in camps and settlements in Bangladesh, Kenya, Tanzania, Iraq and Rwanda. Currently UNHCR is working with the Netherlands Enterprise Agency to develop an open source software platform which has the capacity to visualize the data that is collected by the SWM devices. This software is being developed in an open source collective coding community called GitHub. UNHCR aims to give its partners access to all the required code for this software so that they can use the LoRaWAN hardware to connect their own devices and the software to visualize the data.

Study limitations

The pilot study described was not meant to be a comprehensive or exhaustive test of the range of products on the market but to test the general applicability of SWM technologies in a humanitarian water trucking context. The conclusions of this pilot test are not an endorsement of one technology or product, but rather confirm the value and potential of SWM technologies for humanitarian actors.

Acknowledgments

The authors would like to acknowledge the hard work of the UNHCR staff that participated in the pilot projects in Uganda and Iraq, as well as staff in Bangladesh, Tanzania, Kenya and Rwanda. UNHCR would like to acknowledge the contributions by the Netherlands Enterprise Agency, which supported the development of the open source data visualization platform.

Disclosure statement

No potential conflict of interest was reported by the authors.

ORCID

Ryan W. Schweitzer http://orcid.org/0000-0003-4547-6988

References

Clasen, R., Pruss-Ustun, A., Mathers, C. D., Cumming, O., Cairncross, S., & Colford, J. C. (2014, August). Estimating the impact of unsafe water, sanitation and hygiene on the global burden of disease: Evolving and alternative methods. *Tropical Medicine and International Health*, *19*(8), 884–893. https://doi.org/10.1111/tmi.12330

GBD. (2016). Estimates of the global, regional, and national morbidity, mortality, and aetiologies of diarrhoea in 195 countries: A systematic analysis for the global burden of disease study 2016. *Lancet Infectious Diseases*, *18*(11), 1211–1228. Retrieved November 2018 from https://www.thelancet.com/journals/laninf/article/PIIS1473-3099(18)30362-1/fulltext.

Koolwal, G., & Van de Walle, D. (2013, January). Access to water, women's work, and child outcomes. *Economic Development and Cultural Change*, *61*(2), 369–405. https://doi.org/10.1086/668280

LoRa Alliance. (2015, November). *LoRaWAN: What is it? A technical overview of LoRa and LoRaWAN.* https://www.tuv.com/media/corporate/products_1/electronic_components_and_lasers/TUeV_Rheinland_Overview_LoRa_and_LoRaWANtmp.pdf

Mekki, K., Bajic, E., Chaxel, F., & Meyer, F. (2017, March). A comparative study of LPWAN technologies for large-scale IoT deployment. *ICT Express*, *5*(1), 1–7. https://doi.org/10.1016/j.icte.2017.12.005

OIOS. (2018) *Audit of the operations in Uganda for the office of the United Nations High Commissioner for Refugees* (Report 2018/097 17 October 2019 Assignment No. AR2018/111/01). Office of Internal Oversight Service. https://oios.un.org/ru/file/7247/download?token=48QNfXL1

Sommer, M., Ferron, S., Cavil, S., & House, S. (2015, April). Violence, gender and WASH: Spurring action on a complex, under-documented and sensitive topic. *Environment and Urbanization, 27* (1), 105–116. https://doi.org/10.1177/0956247814564528

UNHCR. (2017). *Status update.*

UNHCR. (2018). *UNHCR management review.* Unpublished internal UNHCR document.

UNHCR. (2019). *Global trends in forced displacement report 2018.* https://www.unhcr.org/5d08d7ee7.pdf

UNHCR. (2020a). *UNHCR WASH monitoring system.* wash.unhcr.org

UNHCR. (2020b) *Global focus insight.* http://reporting.unhcr.org/financial#_ga=2.10537972.194159667.1583087675-115061889.1581009702

UNHCR. (2020c). *Uganda comprehensive refugee response portal.* Retrieved March 6, 2020, from https://data2.unhcr.org/en/country/uga

UNICEF/WHO. (2019). *Progress on household drinking water, sanitation and hygiene 2000-2017. Special focus on inequalities.* UNICEF and World Health Organization.

A GIS-based solution for urban water management

Pablo Fernández Moniz, Jaisiel Santana Almeida, Agustín Trujillo Pino and José Pablo Suárez Rivero

ABSTRACT

Water management systems have an increasing impact on the development of sustainable urban services. The development of a geographic information system applied to water management is presented in this article after taking the priorities of smart cities into account. This system has been fully developed with free open source software to create a main system based on an efficient and specific proposed data model, along with four smart tools that have been developed for the city of Las Palmas de Gran Canaria: water network validations, visual trace simulation, historical data navigation and smart supply cuts. Thanks to this system, the water management company will be able to adapt its policies to offer a better customer experience and consumption plans based on predictive tools. The advantages for managers of water management companies are highlighted and discussed.

Introduction

Turning software into valuable improvement for our planet and giving a clear benefit to our everyday lives is a challenge computers must face in a framework of sustainable development. To address that challenge clearly demands processes such as the development of complex design and modelling questions (how humans interact with tools, data processing and operations) and the construction of coherent spatial relations of the objects represented (data mining and intelligence). Finally, the elevation of all these different abstractions to a human level clearly impacts global sustainability (Sarp et al., 2014), economic growth and climate change.

This article presents a geographical information system–based architecture for smart water management, which is outlined by Fernández, Santana, Sánchez, et al. (2016). Water management companies require different types of software ecosystems to manage a huge variety of information, especially large datasets of interconnected spatial elements: valves, hydrants, pipes, etc. All these elements are usually stored using spatial databases which are used by technicians for analysis, management, planning and monitoring. All the software components that take part in this complex system, where geographical information is also integrated, can be viewed as a geographical information system applied to water management (GIS-WM). Examples of GIS-WMs used around the world are the works of Shamsi (1996) in Pennsylvania (USA), McKinney and Cai (2002) in Central Asia, Calera Belmonte et al. (1999) in La Mancha (Spain), Naphade et al. (2011) in Brazil, and Satti and Jacobs (2004) in Florida (USA).

The main goal of this article is to develop a software solution for the different challenges facing the system being used by the water management company in the city of Las Palmas de Gran Canaria (Canary Islands, Spain). The first challenge is to manage data to permit advanced queries about the water network status. In the past, specific queries could not be answered because the data model was generic and lacked adaptability to the special characteristics of a water network on an island. It also had many unnecessary fields that confused the technicians of the water management company.

Second, the company had serious problems with data quality; for instance, discontinuities in the water network, as well as other topological and alphanumeric issues. This caused data inconsistencies, which generated serious problems in the day-to-day operations of the company.

Third, interoperability problems arose from the software, since not all connections and data loads could be executed unless more module licences were acquired, and that is the fourth problem: licences. Despite not meeting the real needs of the company, the proprietary software requires costly maintenance services.

Several solutions have been developed for some of these challenges individually in the literature. For example, Huang et al. (2010) present a methodology to design a flexible water distribution system that copes with uncertainties and can adapt to changing requirements; Singapore's Public Utilities Board (2016) implemented a smart water grid, integrating information and communications technologies (sensors, meters, digital controls) to ensure that water is efficiently delivered with good quality; and Di Nardo et al. (2017) studied the identification of redundancy features in water distribution networks. More examples can be seen in the book by Christodoulou et al. (2017), where the problem space of assessing system vulnerabilities, failures and risks in urban water distribution networks is fully detailed.

The presented system represents a significant piece of the water management company's policies, including water provision and sustainability decisions. The decision makers who are daily in charge of assuring good service in the city will use information network requirements to improve the quality of the supply and sewage services and establish new policies for the water situation of the island.

More specifically, our proposal is relevant, for example, in planning cuts in water provision. It is critical to ensure that outages affect as few clients as possible, and to guarantee water supply for critical infrastructure, such as hospitals and schools.

Also, in the complex tasks necessary to technically maintain the network, a historical structure allowing inspection of data on previous states of the network is useful. This would allow quick recovery of all the modifications made during network planning and guarantee coherence in the functionality of the entire network, what may be crucial in sustainability decisions.

The structure of the article is as follows. In the next section, the designs of the proposed GIS-WM, data model and software architecture are explained. The third section introduces how the integrity and robustness of the system are ensured using water network validation.

The fourth explains the trace simulation tool that has been developed to assist technicians. The fifth presents the concept of the GIS time machine. The sixth explains the smart tool for automatic and real-time efficient supply cuts in the network. The main guidelines followed for the testing and performance of the system are detailed in the seventh section. The last section presents general contributions and the benefits of using a GIS-WM.

Methods of designing a GIS-WM using free and open source software

free and open source software

As new needs arise in the 'smart city' context, and due to the important role of water management companies there, GISs are no longer isolated and static pieces of software. An accessible data model would pave the way for the development of new analytical tasks to ease the day-to-day work of water management companies. Such tasks would also be adapted to companies' specific requirements.

A GIS-WM should be flexible enough to integrate information sources from internal and external systems by offering database interoperability. The ability to directly control the information should also make it possible to manage raster data, publish on a Web Map Service, and connect to different GIS software solutions to solve different problems.

Instead of providing a simple system to store digitized water networks, a GIS-WM should implement smart capabilities, like helping resolve network errors, while digitizing and helping in the execution of complex analytical geoprocessing tasks. Finally, the system should also support queries of past information and, if necessary, restoring the system to any past date.

Given these considerations, the free and open source software (FOSS) ecosystem offers numerous advantages, including the possibility to use, study, modify, copy and redistribute the source code. This freedom brings several benefits, including avoiding technological lock-in due to restrictive licensing and improving software reusability. Companies also benefit from FOSS because they can reduce their licence expenses and customize their software.

The database management system chosen to implement the GIS-WM is PostgreSQL, with its spatial extension PostGIS. PostGIS is considered one of the most powerful and mature geographic data management software packages in the geospatial market (Akbari & Peikar, 2014).

The software used to connect to PostgreSQL is QGIS (QGIS Project, 2018), a very famous and successful FOSS project in the field of GIS. It is licensed under the GNU General Public License and runs on Linux, Unix, Mac OS X, Windows and Android systems. QGIS has capabilities such as loading data from Open Geospatial Consortium standards, connectivity with multiple database manager system sources, and integration with several geoprocessing tools. It allows adding customized features to address specific user needs with Python plugins. Such features make QGIS a great candidate for the ecosystem of a custom GIS-WM.

Designing the data model

Designing the data model was the first step in creating the GIS-WM. This is considered one of the most important parts of any information system (Codd, 1970). Technologies can change over time, but a well-designed data model remains stable. This is the reason to focus and use the necessary resources to create the new data model. For this study, a tailored data model that meets the requirements of a given company starts from the base of a generic data model. After several iterations with each responsible section in the water management company, we isolate the most important features and discard the generic ones, providing an optimized model for the company (Figure 1).

Table 1 shows the compression rate achieved in a sample of entities shared by the generic and the tailored data models. The new model achieves a significant reduction of the number of attributes used to represent the entities. Therefore, it results in a much lighter and more efficient model tailored to the real needs of the company. The new data model can be easily displayed in QGIS, where the layers are loaded directly from a PostgreSQL database with PostGIS extension (Figure 2).

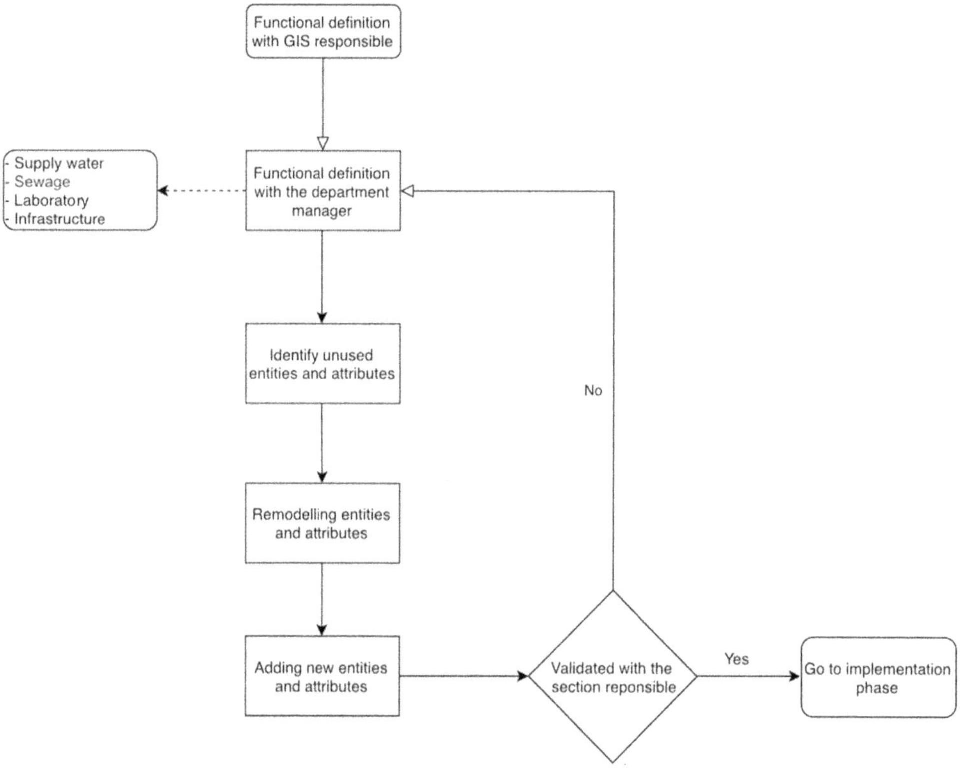

Figure 1. Model design process.

Table 1. Compression rate in the new data model.

	Previous attributes	New attributes	Compression rate (%)
Filters	44	25	56
Connections	47	26	55
Pipes	53	32	60
Enclosures	55	27	49
Valves	62	22	53

Figure 2. Sample network displayed in QGIS client using the customized data model.

Adding flexibility for digitizers: editing and production databases

The daily work of the digitizers involves continuously making changes in the network, but such modifications must satisfy several restrictions and validations to keep the network trustworthy and useful, and to guarantee that users will always query a correct network state. These restrictions and validations are explained in the section on water network validations. To achieve the desired flexibility for digitizers and keep the final network verified, the system was separated into two main databases:

The editing database, where digitizers make the database changes in their daily work. In the editing database, data only persists if it passes a set of critical restrictions defined as Level 1.

The production database, where the 'official' state of water management network is stored, to which direct modifications are not allowed. All elements in this database must satisfy all the restrictions and validations defined for the network.

The way both databases work is the following. An administrator must validate all the changes made by digitizers and import them into the production database. This importing process applies all the restrictions and validations, and only if they pass are all these changes included in the production database. In this way, digitizers can perform their work with enough flexibility, and the network stored in the production database is always verified, what ensures a robust and coherent dataset.

Each database is also connected to another database that stores the complete history of changes. This allows historical information to be retrieved when needed, as explained in more detail in a later section. The general architecture of the system is shown in Figure 3.

Connecting with external systems: assets and commercial management

In this case study there are two important external systems: asset management and commercial management. The asset management system is implemented using an

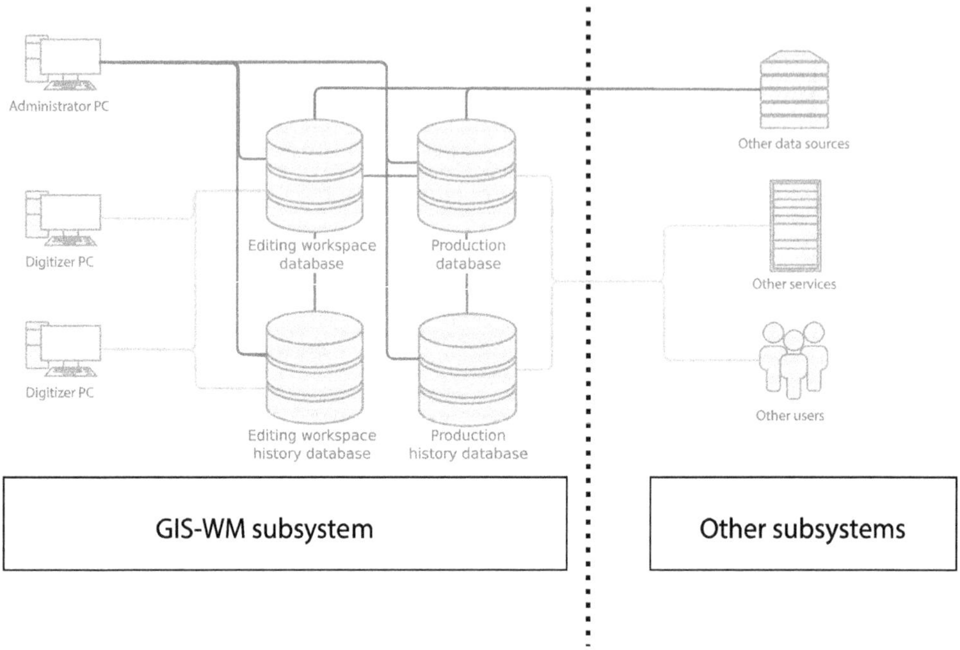

Figure 3. General architecture of the proposed geographical information system for water management (GIS-WM).

Oracle 11g database; it represents an inventory of all the company assets, including production plants, infrastructure and facilities. The commercial management system uses an AS/400 server and keeps all the information related to clients. It is important for companies to connect their different systems together (Robles et al., 2015), and SQL/MED provides a solution to this problem (Melton et al., 2001). The proposed GIS-WM can be extended with customized smart tools. In the following sections, four different smart tools that have been developed are explained in detail.

Water network validation: ensuring integrity and robustness

For a company that offers a very important service like the management of the water supply and wastewater, it is essential to guarantee the integrity and robustness of its network data. To achieve this, a set of rules that verifies that all the defined restrictions are met is introduced. In this section, these validations are described based on geographic operations and the moment they are executed. Finally, a description of the process for performing network modifications is explained.

To categorize the network validations, based on the existence of topological constraints, a validation which involves geometrical or geographic operations while non-topological ones are using them is defined as a restriction. A more detailed categorization can be seen in Figure 4, where non-topological restrictions are subcategorized based on which abstraction level they affect (system, network or entity), while the topological ones are divided by whether they are related to the connection between elements or related to geometry conflicts. Non-topological rules are subcategorized by whether they are applied

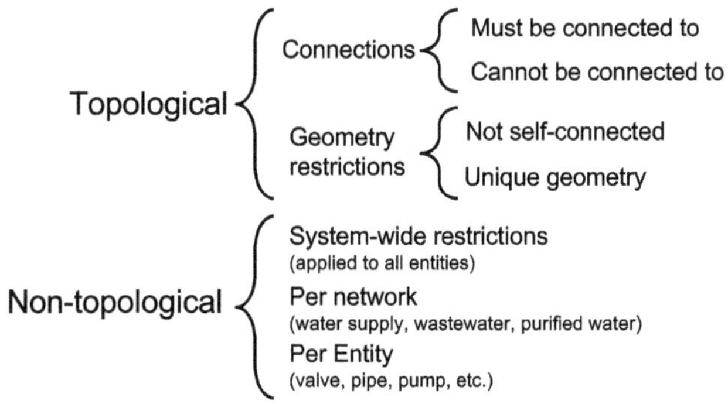

Figure 4. Detailed restriction categorization based on the nature of the validation.

to all system entities, network entities or a single entity type. On the other hand, topological restrictions are grouped by their connections (mandatory or not allowed) and geometry restrictions. In the implementation phase, we tried to generalize the validations as much as possible to simplify the maintenance of the system. This means that if we need to check out a topological property, such as 'lines must be connected', this property is set as generic to be reused in other entities with the same validation, such as connection pipes, which is a different entity in the model from 'standard' pipes.

Currently, there are 77 non-topological restrictions defined in the system. To clarify these types of validations, here is an example of each of the subtypes.

Creation or modification privileges (system level): Only users with digitization privileges can modify the network.

Private network elements cannot be of transport type (network level): Elements that are private property cannot be used to collect water to be stored in water tanks and later used in the general network.

Relationship between diameter and material (entity level): The diameter of asbestos cement pipes must be smaller than 300 mm.

In addition to the non-topological restrictions, there are 172 topological rules. Figure 5 shows two examples, illustrating element connections and geometry restrictions.

Water trace simulation

The day-to-day operations of a water management company involve, among other tasks, the maintenance of the water supply and wastewater networks. To provide good service, it is important to determine the areas and elements affected by pipe breakages, wastewater leaks and so on, so as to plan supply interruptions and deploy the necessary repair solutions.

The water trace tool simulates how water would flow through the network elements based on conditions specified by the user. For instance, the GIS-WM analysts can configure a water trace simulation to stop when a closed valve is found. This allows them to open and close different valves, compare the water traces, and then decide which valves to close before starting any maintenance work.

This tool was developed using two modules: the trace PL/pgSQL function, which does the heavy computation; and a QGIS plugin which serves as the user interface. The water trace simulation tool has the following configuration options:

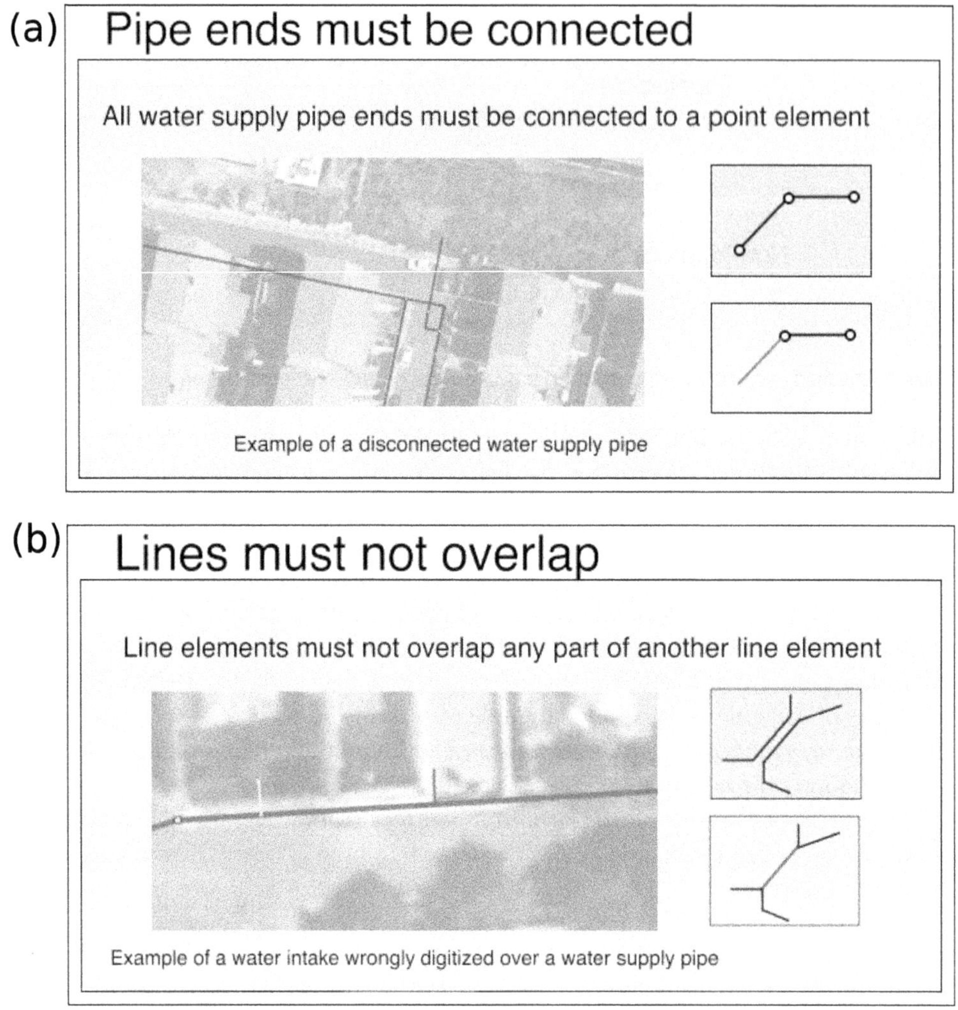

Figure 5. Two different examples of topological restrictions. (a) Proper connection: a point element must be present at each pipe end. (b) Geometry: a line element cannot share a path with another line element.

- *Water flow direction* is a parameter describes how the trace is computed. When its value is 'downstream', the trace will be computed from the start point to the end point in linear entities. For 'upstream', the calculation is reversed. For 'bidirectional', the trace is computed ignoring the direction of the linear entities.
- *Start element* is a key-value structure that represents the network element from which the trace simulation will begin.
- *Stop conditions* a key-value structure that describes at which entities the trace must stop. In this object, every key corresponds with the entity table, and the value corresponds with the query to find the stop element.
- *Display entities* is a list of all the entities to be displayed in the water trace. Even if the water flow simulation runs over all the entities of the system, only those in the list will be returned as the water trace result.

The water trace simulation tool provides a graphical user interface implemented as a QGIS plugin. The plugin guides the user in the process of defining the input arguments described in the tool implementation. At every step, an extensive explanation of the arguments is presented to the user, showing how they affect the water trace simulation to be configured. The plugin also lets the user save the current trace configuration in plaintext files. This allows them to reload the configuration at a future time and avoid reconfiguring tedious or very common traces again. Once the user has defined their desired water trace configuration, they can use the plugin to calculate the traces. The result of the trace is automatically inserted in a separate map layer, so the user can work with it independently. Figure 6 shows how a resulting water trace is displayed over the original network.

Navigating through historical data: a GIS time machine

One of the main requirements in the GIS department of a water company is to provide reports based on the system state at specific dates, to facilitate information exchange internally, or even with external institutions. Since the information required tends to be the same, technicians generate generic reports periodically, but sometimes they need to provide information that was not contemplated in past reports. In these cases, technicians would search through the system backups to restore previous data and generate the needed report, making the process complex and time-consuming for the administrators. This problem was aggravated when the information to be retrieved was present in network states between two system backups.

Figure 6. Trace simulation result using closed valves as stop condition, visualized in QGIS.

In the previous section, the database architecture for the developed GIS-WM was introduced to show that the editing and production databases each have an associated historical database, where changes are recorded. These historical databases include the entity information, the type of operation performed (update, insertion or deletion), and the time range during which this registry was up to date.

To make identifying important events easier, the system implements a milestone table that provides the user an overview of all the important actions performed in the system (data importations and recoveries). In the case of recoveries, this summary also shows the date the system was restored to.

Retrieving historical data

One of the simplest and most used actions performed against historical data is to see what the whole water network looked like in the past. The historical table structure previously described allows users to visualize the state of a network at a past time and copy this past state to the present if desired. This operation could serve as a base for more complicated analytical processes such as comparing different states, detecting mistakes made by a specific digitizer, and generating reports.

The system implements a set of views which facilitate access to a previous system state. Different users do not have to analyze data for the same dates, which means each user needs a custom view. To accomplish this, these views need to filter the historical data by date and by user. This information is retrieved from a system table that stores which timestamp is displayed for each user. Therefore, when the timestamp is changed in that table, the point in time shown to a specific user is also changed.

When technicians detect a possible digitization error related to a water intake imported a couple days earlier, to resolve the issue, they need to visualize a comparison between the actual network state and the network state previous to the importation. The technicians are able to create a set of views of what the system was like before the importation by selecting the timestamp using the plugin.

Restoring previous network states: point-in-time recovery

The point-in-time recovery feature can be used when the administrators of the GIS-WM are not satisfied with the evolution of the network digitization. This feature is presented as data-centred, in contrast with other strategies designed as system backups, like the continuous archiving in PostgreSQL. An important difference from PostgreSQL continuous archiving (PostgreSQL, 2017) is that this feature is not meant to be a backup, and no historical entries are deleted when a point-in-time recovery is executed. On the contrary, the recovery operation is achieved by performing the required inserts, updates and deletions to match the current state to the target state.

The implementation of the historical architecture should be efficient and effective for the storage, visualization and recovery of storage changes. In the developed system, all the generic functions and tables are created automatically based on the database model.

Smart supply cuts

Nowadays having information systems perform smart tasks instead of simply displaying information is becoming more important (Bernard et al., 2014). The GIS-WM introduces the detection of digitization errors and a notification system with hints to solve those errors. This behaviour enhances the digitization workflow, but it is necessary to consider it as a smart feature, a more elaborated information processing.

Using the water trace simulation tool introduced earlier, this system can perform more complex tasks with minimal modifications. Although we do not plan to directly use 'internet of things' data for this example, it should be very useful to enrich the functionality of future tools (Hancke & Silva, 2012; Whitttle et al., 2013).

Planning of supply cuts is a critical operation; technicians must ensure that an outage affects as few clients as possible and maintain the water supply to critical infrastructure, such as hospitals. It is really helpful to have a tool that computes the best network segments to be closed, to restrict the water supply in a specific zone. This tool should help minimize the impact of such operations, taking into account affected clients, valve types, etc.

A simple solution for planning supply cuts has been developed, providing an advanced tool that will help the company improve the quality of service. Using the water trace simulation tool described earlier, it is trivial to show the alternatives for cutting the network by using the valves as stop conditions. Figure 7 shows two alternatives and the different network segments affected.

Once the system has computed the different possibilities, the affected clients are shown for each of them. For that purpose, the system finds all the clients related to the affected

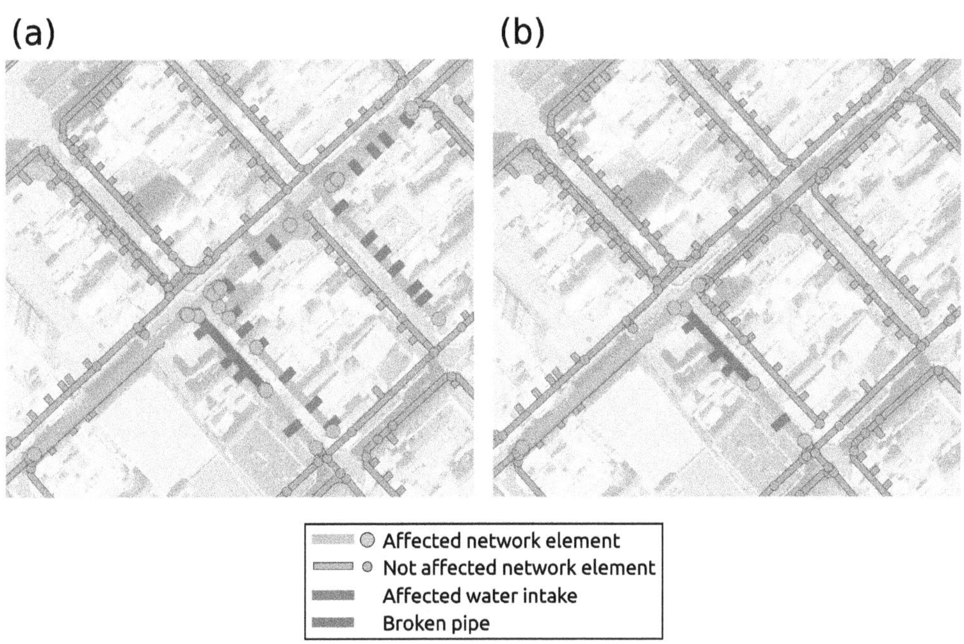

Figure 7. Alternative water supply cuts for the same broken pipe. Both options affect clients with normal priority. Option (b) is preferred over (a) because it affects fewer clients.

water intakes by using the connection to the commercial management system and classifies the traces according to how important it is to guarantee their water supply.

High-priority supply includes traces containing clients that, due to the nature of their activities, cannot be left without water supply, such as hospitals and fire stations.

Medium-priority supply refers to clients that provide less critical services than clients in the previous category, but to which the system must guarantee water supply whenever possible, for example public institutions such as schools and courts.

Normal priority supply refers to clients that do not need any special treatment.

With all the information gathered, the system proceeds to evaluate which is the best option. For this, a simple rule-based system was designed and written in CLIPS (Giarratano & Riley, 2005). The rule-based system selects the best places to close the network based on the categorized traces, the number of clients affected and other properties of the trace. When a rule is met by two different cut paths, the system marks the best path and does not evaluate the rest of the rules for that pair. After all the pairs of alternative cut paths are compared, the system decides which is the overall best path to cut the water supply. The rules considered by the rule-based system are (in decreasing order of importance):

Water supply priority: select the network segments that only affect clients with the lowest priority of supply.

Number of affected clients: mark a trace as better than another if it affects fewer clients.

Number of valves to be operated and their types: a path with fewer valves to be operated is better. On the other hand, remote-control valves are preferred over manually activated valves.

With these rules, if there are two options for the supply cut, and both of them only affect clients without special needs related to water supply, the system will choose the trace that affects fewer clients. The assistant is very simple, but it could be easily improved by including more characteristics and writing additional and more complex rules. Here are some examples.

Expected drain time: network segments that drain the water in a shorter time could be prioritized.

Distance between valves: the shorter the distance a technician must travel to operate all the valves, the better.

Work being performed in the network: check for operations currently being performed in the network. This rule would need to use the connection with the asset management system (explained in the section on connecting with external systems).

Testing and performance

One of the biggest challenges in the modelling process consists of testing whether the final product is close to the requirements, and provides a real improvement for the company. That's why throughout the definition process we used a continuous control system and internal trainings to synchronize with the water management staff.

Survey of satisfaction before and after the project

During the first phase of the project, several iterations were made with the persons responsible in each department of the company. As was explained in the earlier section, during the definition process we made several iterations to ensure alignment with the water management company. The main communication process was managed orally and with minutes of the meetings.

Moreover, in the final phase of the definition of requirements, several records were created to ensure synchronization with needs and requirements, from simple requirements such as permissions for each different role in the system (Figure 8), to defining the components and validations for each entity.

A satisfaction survey was also conducted to measure improvements made for the next iteration once the system was used. The survey covered the whole GIS department and the different sections of the company, a total of 18 participants. This gives a sense of the general level of satisfaction with the implemented system (Table 2).

Technical performance

One of the main problems we found in the creation of this water management tool were the validations. Validations are a good thing to have in a conceptual or a new system. But a very strict system of validations means that all of the network must satisfy all of the rules. This was a big problem in the implementation process. The water management company promised that the whole network should be ready and accomplished once the development was finished, they could not meet this requirement.

Data Bases	System	Admin	Digitizer	Read user
CREATE	V	X	X	X
TEMP	V	X	X	X
CONNECT	V	V	X	X

Shema	System	Admin	Digitizer	Read user
USAGE	V	V	X	X
CREATE	V	V	X	X

Tables	System	Admin	Digitizer	Read user
INSERT	V	V	V	X
SELECT	V	V	V	V
UPDATE	V	V	V	X
DELETE	V	V	V	X
TRUNCATE	V	V	X	X
REFERENCES	V	V	X	X
TRIGGER	V	V	X	X

Figure 8. Defining permissions.

Table 2. Results of the survey.

Closed questions (answers from 1 to 5):	
Evaluate in a general way the improvements of the new tool	4.3
Evaluate the loading timing compared to the previous tool	5
Evaluate the improvements in day-to-day work and specific tools	3.9
Open question (% of respondents):	
Which parts of the system need improvement?	
None	83.3
It would be great to have advanced line-node editing tools	11.1
No answer	5.5

To deal with this problem, we added a new flag field to the system. This field allows (but flags) elements that are in a wrong state, such as a pipe that is disconnected. The next time a digitizer imports or validates something in this zone, an alert is given. This makes it easier to create new elements, and at the same time, it helps in editing the old elements that do not pass the validations in a progressive way.

More general testing was carried out during the development process, including:

- Concurrency tests,
- Load tests,
- Unit testing of each function of the system, and
- Functional test by replicating all the workflow in the system.

General contributions and benefits of the GIS-WM

The GIS-WM system described in previous sections has many advantages for managers of water management companies and their technicians. The most important advantages are:

- *Better software costs investment.* Companies can save money on licence costs and invest it in further improving their systems.
- *No licence-per-user limitation.* The number of people authorized to work on digitization tasks is not limited by the number of licences purchased. With a limited number of concurrent users, companies cannot add new digitizers to help with map editing during peaks in workload. And if new licences were acquired, they would need to be tailored for the specific dates when the new digitizers would work; otherwise, the company is wasting money on unused licences. With the new system, the number of people able to connect to the system is dictated only by technical limitations and management decisions, eliminating purely economic limitations due to costs per user (Figure 9). The GIS-WM has a clear economic benefit in the medium term.
- *Control over software extensions.* This solution not only connects with other vendors' software but it is also open to extend support for other subsystems and technologies. Therefore, companies can choose what to extend and who should implement those extensions.
- *Higher network digitizing quality.* The complete validation system improves the digitization workflow, raising digitizers' productivity and the speed and quality of network error correction.

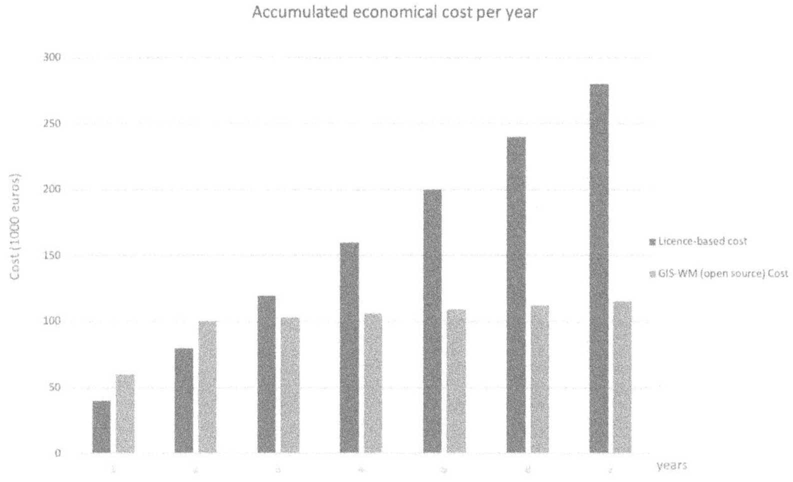

Figure 9. Cumulative cost per year, comparing the GIS-WM software and other licensed software in the same company.

- *Better backup management.* The historical management system presented in this article allows administrators to revert the database state to any desired time instead of a state saved in a periodic backup.

In addition to all of these benefits, the system allows the company to develop more advanced features to improve its services and work efficiency, like the supply cut planning assistant described earlier. Other assistants that could be developed using the architecture of the system and the connections with other data systems of the company are:

- *Breakdown detection.* This assistant would continuously read sensor data, and when an unexpected reading is found, notify the administrator. The unexpected reading could be, for example, a difference (greater than some threshold) between the volume of water entering a network segment and the water leaving it, or excessive chlorine in the water. Afterwards, the assistant could indicate where the fault could have been occurred by performing a trace against the water flow, using as the stop condition the network elements where measures were correct. With the elements obtained from the trace, and taking into account the fault type, the system could suggest which elements should be checked first, or ask for complementary measures. For example, if work has been performed recently near the affected trace, the system could mark the elements that were worked on as a possible cause.
- *Network replacement and maintenance.* This assistant would suggest which network sectors should be replaced or need maintenance. Using data-mining techniques, the system would search for patterns connected with network renovation. An example is the identification of a defective pipe lot if an abnormal number of pipes have replaced before expected end of life. Another pattern could be to calculate a more accurate life cycle based on the pipe material, the water pressure and water composition parameters such as hardness.

- *Web-based assistants.* Web applications where the GIS-WM information is integrated offer a great range of possibilities. Citizens might be able to check whether their house is affected by a supply cut or fault. This service could be easily developed using the smart assistant for supply cut planning described earlier and to access data on which clients were affected at a later stage. Another example would be the creation of a dashboard for technicians, such as the Smartwater application (Figure 10), which was presented at the Smart City Expo World Congress in Barcelona. This application uses FIWARE technology to collect water quality measures from sensors. The collected data is crossed with geagraphical information, and then presented to water network analysts in a web viewer. The main aim is to provide a decision support system, like the one available in SmartPort (Fernández, Santana, Ortega et al., 2016), with an alert system providing the values given by sensors, to help operators control everything related to the water ecosystem.

Conclusions

We have described the development of a GIS applied to water management (GIS-WM) to deal with the challenges of the water management company in the city of Las Palmas de Gran Canaria (Canary Islands, Spain). This system is presented as a tool for strategic decision making and for setting up strategies for the water management company in accord with its policies. Taking into account the priorities of smart cities, the system has been designed to guarantee interoperability between systems, facilitating the day-to-day work of the water network digitizers, and supporting the development of advanced smart analytical processes. Its implementation has been possible thanks to the use of FOSS, diverse data and communication standards, and the creation of a

Figure 10. Retrieving sensor data with the Smartwater application. The blue lines represent pipes in the water network.

customized data model to cover specific features. Adopting FOSS for the GIS-WM provides a very competitive solution in terms of cost in comparison to other commercial software solutions.

The decision to separate the system into two main databases allows the digitizers to freely work in a working environment, while the validation system ensures that whenever the changes are ready, they are safely stored in the production environment. This validation system guides the digitizers through all the needed changes, which improves their working experience and the quality of the digitized water network. The design of the architecture and the processes minimize the number of validations executed, considerably reducing the importing time.

The GIS-WM also provides a historical management system to store all data changes made in the digitized water network. This provides the ability to inspect the state of the network at any desired date. Having all the modifications of the system accessible enables studying the evolution of the data. This brings flexibility to administrators and managers, who can choose to recover the system to any past state without losing the history after the recovery, in contrast with common backup systems. Different implementations, using PostgreSQL and MongoDB, were studied to decide which was the best for this historical system.

The ability to directly access the data model and the use of standards that facilitate interoperability between systems allow the integration and development of diverse tools, such as the water trace simulation tool presented in this article. The ease of integration with different smart assistants, like the supply cut smart assistant, has also been shown. Other potential tools using the existing resources have also been described.

The proposed GIS-WM would give companies not only economic but also operational advantages. And all of this, along with the flexibility to improve and expand their implementation to better support their activities, directly affects the sustainable development of cities. Upcoming challenges for the GIS-WM are adding more automation to perform operational tasks and provide smarter system behaviour; adding more data-related utilities, like statistics and network performance; and including smartphones and tablets as devices to enter and edit data, e.g., for digitizing purposes.

Disclosure statement

No potential conflict of interest was reported by the authors.

ORCID

Pablo Fernández Moniz http://orcid.org/0000-0002-9443-9043
Agustín Trujillo Pino http://orcid.org/0000-0001-6212-5317
José Pablo Suárez Rivero http://orcid.org/0000-0001-8140-9008

References

Akbari, M., & Peikar, S. R. H. (2014). Evaluation of free/open source software using OSMM model case study: WebGIS and spatial database. *Advances in Computer Science*, *3*(5), 34–43.

Bernard, T., Piller, O., Gilbert, D., Braun, M., Kühnert, C., Deuerlein, J., ... Werey, C. (2014, August). *SMaRT-OnlineWDN: A Franco-German project for the online security management of water distribution networks*. Paper presented at the 11th International Conference on Hydroinformatics, City University of New York.

Calera Belmonte, A., Medrano Gonzalez, J., Vela Mayorga, A., & Castaño, S. (1999). GIS tools applied to the sustainable management of water resources: Application to the aquifer system 08-29. *Agricultural Water Management*, *40*(2), 207–220. https://doi.org/10.1016/S0378-3774(98)00122-X

Christodoulou, S., Fragiadakis, M., Agathokleous, A., & Xanthos, S. (2017). *Urban water distribution networks: Assessing systems vulnerabilities, failures, and risks*. Butterworth-Heinemann.

Codd, E. (1970). A relational model of data for large shared data banks. *Communications of the ACM*, *13*(6), 377–387. https://doi.org/10.1007/978-3-642-48354-7_4

Di Nardo, A., Di Natale, M., Giudicianni, C., Musmarra, D., Varela, J. R., Santonastaso, G. F., Simone, A., & Tzatchkov, V. (2017). Redundancy features of water distribution systems. *Procedia Engineering*, *186*(2017), 412–419. https://doi.org/10.1016/j.proeng.2017.03.244

Fernández, P., Santana, J., Sánchez, A., Trujillo, A., Domínguez, C., & Suárez, J. P. (2016). A GIS water management system using free and open source software. In C. R. García, P. Caballero-Gil, M. Burmester, & A. Quesada-Arencibia (Eds.), *Ubiquitous computing and ambient intelligence* (pp. 383–394). Springer International.

Fernández, P., Santana, J. M., Ortega, S., Trujillo-Pino, A., Suarez, J., Dominguez Trujillo, C., Sánchez Medina, A., & Santana, J. (2016). SmartPort: A platform for sensor data monitoring in a seaport based on FIWARE. *Sensors*, *16*(3), 417. https://doi.org/10.3390/s16030417

Giarratano, J., & Riley, D. G. (2005). *Expert systems: Principles and programming* (4th revised ed.). Course Technology Inc.

Hancke, G., & Silva, B. (2012). The role of advanced sensing in smart cities. *Sensors (Basel, Switzerland)*, *13*(1), 393–425. https://doi.org/10.3390/s130100393

Huang, D., Vairavamoorthy, K., & Tsegaye, S. (2010). Flexible design of urban water distribution networks. In *World environmental and water resources congress 2010: Challenges of change* (pp. 4225–4236). American Society of Civil Engineers. https://doi.org/10.1061/41114(371)430

McKinney, D., & Cai, X. (2002). Linking GIS and water resources management models: An object-oriented method. *Environmental Modelling and Software*, *17*(5), 413–425. https://doi.org/10.1016/S1364-8152(02)00015-4

Melton, J., Michels, J.-E., Josifovski, V., Kulkarni, G., Schwarz, K., & Zeidenstein, K. (2001). SQL and management of external data. *ACM SIGMOD Record*, *30*(1), 70–77. https://doi.org/10.1145/373626.373709

Naphade, M., Banavar, G., Harrison, C., Paraszczak, J., & Morris, R. (2011). Smarter cities and their innovation challenges. *IEEE Computer*, *44*(6), 32–39. https://doi.org/10.1109/MC.2011.187

PostgreSQL. (2017). Continuous archiving. PostgreSQL. https://www.postgresql.org/docs/9.6/static/continuous-archiving.html

Public Utilities Board Singapore. (2016). Managing the water distribution network with a Smart Water Grid. *Smart Water*, *1*, 1–13. https://doi.org/10.1186/s40713-016-0004-4

QGIS Project. (2018). *QGIS*. http://qgis.org

Robles, T., Alcarria, R., Martín, D., Navarro, M., Calero, R., Iglesias, S., & López, M. (2015). An IOT-based reference architecture for smart water management processes. *Journal of Wireless Mobile Networks, Ubiquitous Computing, and Dependable Applications*, *6*(1), 4–23.

Sarp, S., Won Lee, S., Jeon, D. J., & Kim, J. H. (2014). Smart water grid: The future water management platform. *Desalination and Water Treatment*, *55*(2), 339–346. https://doi.org/10.1080/19443994.2014.917887

Satti, S., & Jacobs, J. (2004). A GIS-based model to estimate the regionally distributed drought water demand. *Agricultural Water Management*, *66*(1), 1–13. https://doi.org/10.1016/j.agwat.2003.10.003

Shamsi, U. M. (1996). Storm-water management implementation through modeling and GIS. *Journal of Water Resources Planning and Management*, *122*(2), 114–127. https://doi.org/10.1061/(ASCE)0733-9496(1996)122:2(114)

Whitttle, A., Allen, M., Preis, A., & Iqbal, M. (2013). *Sensor networks for monitoring and control of water distribution systems*. MIT Open Access Articles.

SWM and urban water: Smart management for an absurd system?

M. P. Trudeau

ABSTRACT

Only a very small fraction of the potable water produced by urban centres is used for a need requiring the highest-quality water. Urban water infrastructure has evolved over the past century without considering whether past practices make sense for a future with stressed ecosystems and a changing climate. Smart water techniques are vital to optimize existing infrastructure. However, for urban water servicing of the future, guiding principles developed through consultation, a long-term vision, and tailor-made plans for local conditions are needed. A coordinated research agenda to address many common challenges could support implementation of the plans and the vision.

Supervisory control and data acquisition systems help optimize the infrastructure that reliably delivers high-quality drinking water to urban dwellers in many cities around the world. These advanced systems optimize pumping and energy use. They also monitor reservoir levels to meet expected demands. However, according to an estimate by the American Water Works Association, just a small fraction (Deb & Evans, 1980) of the potable water produced by modern urban water infrastructure is used for a need that is matched to its quality. While monitoring pollutants to the parts per billion in drinking water, we are using that same water to fight fires, to wash cars, to water lawns, and in industrial processes. For truly smart water management, in addition to optimizing current systems, we need a 100-year vision for a fundamental realignment of urban water infrastructure.

That only a very small fraction of potable water is used for purpose is surprising at first glance, but it is the product of a progression of logical decisions made over the past 100 years that has developed into an illogical system. It is time to look back on those decisions with a view to charting a new direction. Future conditions and constraints that are anticipated from climate change and increasing human population concentrations in urban centres will no longer support systemic inefficiencies in urban water infrastructure. To develop solutions, a strategic perspective is needed for the entirety of urban water infrastructure, including potable, wastewater and stormwater components. This infrastructure of the future must respect the water cycle and the aquatic life forms it supports. With a strategic understanding, a set of guiding principles can be proposed to avoid the indefinite perpetuation of designs and operational standards rooted in centuries past.

How did we get here?

To understand how potable water infrastructure evolved to such a condition of fundamental structural inefficiency, it is helpful to go back in history to a time when acute threats to human health and safety were the primary concern of communities served by these modern systems. The system in Ottawa, Canada, provides an excellent case study for tracking the development of potable water infrastructure since it parallels similar systems in North America, Europe and elsewhere.

In 1870, a massive fire threatened the downtown core of Ottawa after having burned several hundred square kilometres of farmland and villages in the outskirts, leaving thousands homeless and killing about 20 people (Powell, 2014). The fire was stopped before it could reach the downtown core by the release of water from a dam, allowing a torrent of water to form a barrier to the flames (Powell, 2014). This event, and the Great Chicago Fire in 1871, prompted Ottawa's municipal politicians to finally overcome the resistance of private water carriers and approve the construction of a pumping station to supply water to the city (City of Ottawa, 2001). The Fleet Street Pumping Station drew water from the Ottawa River for distribution throughout the urban core. The pumps in the station were – and still are – powered by water diverted from the river to run turbines using hydropower.

At the time of its completion in 1873 (City of Ottawa, 2001), this station provided fire suppression capacity using untreated water from the Ottawa River. The water pumped from Fleet Street was also a convenient source of water for domestic, industrial, commercial and institutional uses. Since that time, engineering standards have advanced to ensure that water for fire suppression is available in sufficient quantities at defined minimum water pressures, even under unusual circumstances. For instance, the Canadian Fire Underwriters Survey stipulates that residential fire flows should not be less than 4800 litres per minute for two hours, plus additional capacity to meet maximum daily domestic demand; commercial, institutional and industrial areas must be serviced at minimum fire flows of 5000 litres per minute (City of Toronto, 2009). The United States applies the International Standards Organization insurance criteria (Hickey, 2008) which indirectly influence property values, property tax levies and insurance premiums, depending on the fire suppression rating schedule (Weber, 2018). In designing potable water distribution systems, back-up power generation capacity is provided at key pumping stations in case a fire occurs at a time when the normal power source is not available. Standards also exist for fire hydrant spacing, water main sizing, and water production and storage capacity to meet potential fire suppression demands. The distribution system in Ottawa is designed to meet peak water demand by customers plus fire demand.

In 1911, a typhoid epidemic broke out in Ottawa (City of Ottawa, 2001). In England in 1854, Dr John Snow had established a link between contaminated water from the Broad Street Pump in London and cholera deaths (Benidickson, 2007). The belief that rivers contaminated with sewage could self-purify was defeated by the 1880s (Benidickson, 2007). Until these scientific findings, water quality concerns were of secondary importance due to a lack of understanding of the spread of disease from water sources (Benidickson, 2007). Once this was understood, advancements in water purification techniques received increased attention. Chlorine was recognized as a chemical that could eradicate bacterial infections, and in 1911 chlorine (in the form of hypochlorite) was added to the water distributed by Ottawa's Fleet Street Pumping Station (City of Ottawa, 2001). However, chlorine must make direct contact with the bacteria, so it is not effective in water that has a high concentration of sediments which can shelter the bacteria from contact with the chemical. A pipe failure caused contamination of Ottawa's water supply in 1912, and a second typhoid outbreak occurred in that year. These outbreaks prompted funding approval for more advanced purification technology to be implemented (City of Ottawa, 2001).

Through the science of water purification, the sand filtration technology still in wide use today was developed. This technology entails pumping water onto engineered layers of anthracite, sand and gravel to filter out contaminants. Chemicals called coagulants are added before the filter system to water that flows very slowly through settling basins, allowing smaller particles to clump together, making them too large to pass through the engineered filter. In Ottawa, this advanced treatment technology could not be implemented at the Fleet Street Pumping Station, so a new facility was constructed to purify water that was then fed into the pumping station for pressurized distribution. In 1931, the Lemieux Island Water Purification Plant was commissioned (City of Ottawa, 2001). Even today, purified water from the plant is conveyed to Fleet Street, where it is pressurized and distributed to the city.

With the resolution of the acute health risk posed by bacteria, scientific research began to identify chronic human health risks from drinking water. For instance, when chlorine comes into contact with organic materials in untreated water, by-products form, and these hold the potential for increased risk of cancer. These disinfection by-products, such as trihalomethanes (THMs) and haloacetic acids (HAAs), can be reduced by modifying the purification process to delay the application of chlorine until after filtration. These modifications and others continue today to reduce the risks of chronic disease from long-term exposure to drinking water from municipal water purification plants. The ability to detect and monitor contaminants in drinking water has led to the standardization of allowable limits of contaminant concentrations. The World Health Organization (2017) lists approximately 90 constituents of potential concern in drinking water, some with allowable limits in the parts per billion.

Mismatched objectives for potable water infrastructure

With this brief history, the underlying structural design flaw is readily apparent: water is purified to meet the quality standards for human consumption, and then it is pumped into a distribution system designed for fire suppression. These two fundamentally different functions are not delivered independently. Their conflation into one urban system results in the shocking reality that only a small fraction of the water produced is used for a need that requires the highest water-quality standard. Even within homes, where the direct use of highly purified water prevails, only about 20% of the water used issues from faucets for drinking, and another 20% for bathing. The remaining 60% is used outdoors, for toilet flushing and for clothes washing (Bipartisan Policy Center, 2017). Depending on the economic profile of a community, industrial and commercial uses may demand over half the potable water produced for a municipality, in some cases with little need for highly purified water.

Smart water management techniques have been, and are being, developed to optimize purification processes, monitor contaminants, ensure adequate system storage, reduce energy use in distribution systems, and more. However, the design of the system precludes addressing the most blatant source of inefficiency: *fire suppression does not require drinking-quality water.* The minimum-flow requirements for fire suppression bring water to each door, enabling multiple other uses and encouraging uses that do not require high-quality water.

There are other statistics that cause the uninitiated to wonder at the apparent lack of concern on the part of water professionals for systemic inefficiencies. For instance, at least 10% of the water produced for drinking is lost in the distribution system through leaks, water theft, and inaccurate meters that under-record consumption by customers. In fact, 10% is a rule-of-thumb objective for water lost from production to consumption in a municipal system. If less than 10% of the potable water produced is lost, there is a possibility the municipality or water utility is spending more money on repairs than is warranted by the potential revenues lost from not being able to sell the lost water to customers. This calculation for water loss is a product of a global economic system that does not place an intrinsic value on natural resources but rather tracks the costs of activities to harvest and use natural resources. Although 10% of the volume produced is a lot of water, the cost of finding and repairing leaking distribution pipes can be prohibitive for their repair.

Smart technologies have been developed to find leaks in buried water mains, such as sonic devices that can 'hear' water under pressure escaping from pipes. The question becomes: Is the cost of producing that water higher than the cost of stopping the leak? Consider that water mains must be below the depth to which frost penetrates under the roadway. In Canada and parts of the northern United States, the frost line can be over two metres deep (in Ottawa, it is three metres). To repair a leak or to replace an old water main, the asphalt or concrete roadway must be removed and a trench dug down to the pipe, which is typically underneath the infrastructure of other utilities, providing gas, telecommunications, and electrical power to communities. The sanitary sewer and storm sewers may also need to be circumvented, although these are often deeper than the purified water system pipes. Representatives from other utilities must provide location information for their service lines, traffic has to be diverted, heavy equipment brought in for the excavation, and proper safety measures used at all times, such as an excavation trench box to prevent suffocation of the workers if the earthen sides of the trench collapse.

Considering the cost and disruption needed to repair leaking water mains, the costs of the chemicals and pumping energy invested in the escaping water are usually not the deciding factor to excavate. Rather, the risk of catastrophic failure of the roadway from water erosion, loss of system pressure, the condition of the roadway, and other similar considerations are the factors that inform decisions to repair, replace or ignore leaking water mains.

Maintaining high water quality in potable water distribution systems requires continuous demand to prevent stagnation. A residual concentration of chlorine is needed for all potable water in the distribution system to ensure that pathogens cannot thrive. If water sits too long in a pipe, the chlorine residual can be lost, and a risk of contamination of the water arises. With the pipes sized for fire supply, large volumes of water are available even when no fires are being fought. A continuous demand for water helps keep the quality protective of human health. Thus, drastic reductions in potable water use could ultimately backfire if water stagnates in the potable water distribution system.

It was logical to pump water into the city to fight fires. Because that water was also used for domestic uses, it was logical to add chlorine to prevent acute health threats. As health risks became better understood, it was logical to improve purification processes. As cities expanded, it was logical to extend the existing water distribution system to new subdivisions. But, at some point in this history, the system became absurd: potable-grade water is used for multiple applications that do not require that level of quality. It is not the purpose of this article to lay blame or to identify a point in time where different decisions could have been made. The point is that we now need to look at the reality of urban water infrastructure to formulate a vision for systems of the future that are not absurd. However, to formulate this vision, we cannot look at the drinking water system in isolation.

A second component of urban water infrastructure: wastewater

Drinking water infrastructure influences sanitary sewer design. Specifically, the slope of the sanitary sewer is designed for a minimum scour velocity, or in other words, the slope is designed to maintain a flow rate that will carry sewage through the sewer to a wastewater treatment facility, assuming it is not discharged directly to a water body. This scour velocity is based on assumptions about how much water will flow through the sewer, which in turn depends on how many showers, toilet flushes, washing loads, industrial processes and commercial activities send water to the sewer. A dramatic reduction in water entering a sewer system from the potable water system would necessitate sewer pipes with a greater slope and more repumping of sewage to ensure that the material is properly conveyed to the wastewater treatment facility.

More fundamentally, from a design perspective, the assumption that human waste belongs in water has long merited examination. As with the potable water system, a series of very logical decisions have gone into the creation of modern urban sanitary sewers. An unspoken assumption underpinning sanitary sewer system design is that human waste should be discharged to water bodies. This age-old practice dates back to early urban settlements, when sewers were designed to carry sewage and street runoff directly to watercourses (Benidickson, 2007), but we are only now coming to an understanding of the breadth of the implications for humans and aquatic ecosystems.

The typical wastewater treatment facility uses screening and settling (primary treatment) with biological treatment (secondary treatment) to reduce suspended solids, oxygen demand and pathogens in the effluents released to the freshwater environment. A primary objective of treatment is to protect aquatic life by preventing harmful low oxygen concentrations in the receiving waters. Wastewater treatment plants also reduce suspended solids concentrations and, in some cases, are required to augment the reduction of nutrients (i.e. phosphorus and nitrogen) and to disinfect for pathogens in their effluents prior to release.

While treatment systems improved the water oxygen levels available for aquatic life downstream of sewage outfalls, the role of nutrients was not initially understood or considered. Phosphorus was transferred from agricultural fertilizer to farm produce to digested foods to sewers to watercourse without appreciation for the fact that this nutrient stimulates biotic growth in freshwater environments. Reconsideration of wastewater treatment processes was prompted in the 1970s with the confirmation that phosphorus was a limiting nutrient in freshwater systems, causing eutrophication in sensitive ecosystems (Schindler et al., 2012). More recently, attention has turned to other contaminants in wastewater effluents, including metals, toxic by-products of disinfection, pharmaceuticals, micro-plastics, pesticides and hormone-mimicking substances. The effects of these wastes remain poorly understood in terms of their long-term effects on freshwater ecosystems. For instance, substances that interrupt hormones in fish populations result in feminization of fish downstream of urban wastewater treatment plants in Canada and elsewhere (Hamid & Eskicioglu, 2012; Kidd et al., 2007); but the long-term implications for ecosystems are unknown (Marcogliese et al., 2015). New contaminants continue to emerge, in particular from nanotechnologies and industrial chemicals (Shen et al., 2016).

The wastewater treatment technologies installed in most cities date back to the mid-twentieth century. These systems were never designed to remove pharmaceuticals and other chemicals generated by our current lifestyles. Elements such as cadmium, chromium and lead are not magically transformed, nor do they disappear. They are only 'treated' by wastewater systems in the sense that some portion is removed from the water phase and transferred to solids. The solids, which result from settling and the secondary treatment process (called biosolids), contain metals and other contaminants, but they also are potentially an important source of nutrients. The appropriate use or disposal of biosolids is currently a raging debate that will not be explored here (see Collivignarelli et al., 2019). Suffice it to say that the phosphorus in biosolids has the potential to offset the dwindling reserves of easily accessible mined phosphorus, while also reducing the burden of this nutrient's release to fresh water.

The optimum solution to toxic materials in wastewater would be to prevent them from entering sewers in the first place. The term 'source controls' refers to methods to prevent the release of undesirable substances by stopping them at their source of introduction to a waste stream. Such controls may help reduce or eliminate metals and toxic substances released from industrial processes or by the use of consumer products. Specialized wastewater treatment may be imposed before the release of wastewater from industrial processes to a municipal sewer, but strong political will is needed to work with corporations to ensure compliance with more stringent sewage quality requirements. However, source controls for pharmaceuticals are not currently an effective means of reducing the release of these chemicals into sewer systems. These substances are released primarily in the human sewage of people who have consumed the pharmaceuticals for health or other reasons. Source control in these cases entails redesigning toilets and/or reconfiguring sewage collection systems so that the pharmaceuticals can be collected before their release to, ultimately, a water environment. This alternative approach also enables direct access to nutrients through 'ecological sanitation' (Höglund, 2001; Swedish International Development Cooperation Agency & Esrey, 2001).

Full circle

As the consequences of using water to dispose of human waste become increasingly apparent, the true nature of the water cycle has been revealed: what goes into water comes back around in water. There are no quick or cheap ways to remove trace contaminants in potable water or wastewater using conventional municipal infrastructure, either through wastewater treatment or through water purification processes. Many governments are not investing sufficient funds to maintain the existing systems (American Society of Civil Engineers, 2016; Canadian Infrastructure Report Card, 2019), let alone being in a position to consider upgrading to advanced treatment technologies for potable or wastewater plants. In addition, the health effects of many of these contaminants are unknown or do not warrant the high investment costs of installing more advanced purification or wastewater treatment processes, such as ultrafiltration membrane techniques, which have their own water demands and by-product wastes. Given the structural inefficiencies and unintended consequences of current urban water infrastructure configurations, wholesale reconsideration of the systems is warranted before the commitment of the many billions of dollars that would be required for technological upgrades to plants.

There are two components remaining for consideration before discussing potential solutions for a path forward: stormwater and the natural system itself.

A third piece of the urban water puzzle: stormwater

Stormwater infrastructure has not traditionally influenced potable water infrastructure design, but it certainly could play a role. In Ottawa, the peak demand for potable water occurs in early June (City of Ottawa, 2013). Gardeners, green space managers and homeowners, who are keen to ensure that plants get a good start at the beginning of the short growing season, plant during the first weeks after the risk of overnight frost has passed. This risk has passed by early June, and thus outdoor water use for gardening begins. The maximum day water use in Ottawa in 2012 was 1.6 times higher than the base day demand registered during winter months (City of Ottawa, 2013). What does this peak have to do with stormwater? Conventional urban stormwater infrastructure in Canada and elsewhere does not store rainwater for later use. Many of the current applications for which potable water is used could be served by rainwater, for example, gardening, street sweeping, car washing or firefighting. Green infrastructure and low-impact techniques (CVCA & TRCA, 2010) call for cisterns or other mechanisms to store rainwater, but these approaches are far from commonplace, even in new subdivisions. The limited storage of rainwater for irrigation during dry periods prompts the use of highly refined potable water on outdoor gardens and lawns.

As more land is taken up for urban developments, the consequences for aquatic ecosystems continue to mount. Contaminants from roadways, such as windshield washer fluid, brake fluid, metals from brake linings, salt from winter road operations and more, are washed directly into receiving creeks, rivers and lakes. Rudimentary treatment systems for stormwater only started to be installed in earnest in the 1980s, and these are typically passive systems that rely on contaminants settling out when the water slows down. Regular maintenance is required to clean out the settled material to prevent it from washing out of the stormwater facilities during the next high-flow rain event.

Brief overview of urban aquatic ecosystem conditions

The original design objective for stormwater systems, once they were separated from the sanitary sewer network, was to whisk water off streets and properties as quickly as possible, leaving no puddles behind (CVCA & TRCA, 2010). The implications for receiving watercourses were not considered. One direct and obvious result of stormwater discharge is the erosion of stream banks, but the scope of change is much more extensive. The increased volume of runoff from impervious surfaces (i.e., streets, sidewalks, roofs) that do not allow infiltration of water imposes substantial alterations on the flow regime (Trudeau & Richardson, 2016) and water chemistry, sediment transport regime, water temperature, and other aquatic habitat features.

Land use change is a recognized stressor for aquatic biodiversity (Klein, 1979; Löfvenhaft et al., 2004; Paul & Meyer, 2001; Schueler, 1994; Stanfield et al., 2006; Stanfield & Kilgour, 2006). As the devastating effects of urbanization on watercourses began to emerge (Adamowski & Prokoph, 2013; Boers et al., 1986; Brabec et al., 2002; Costa et al., 2003; Dudgeon et al., 2006), the term 'urban stream syndrome' (Walsh et al., 2005) was coined in the scientific literature in recognition of the numerous concurrent negative effects of stormwater effluents on urban waterways.

In an analysis of the relationship of aquatic biodiversity and water security, Vörösmarty et al. (2010) found that threats to aquatic biodiversity increase with increased water security due to the infrastructure built to sustain that security, including dams and reservoirs. These systems alter water flows, chemistry and temperature. The changes in flow regime have serious consequences for fish diversity in urban streams (Trudeau & Morin, 2017). Aquatic community health is anomalous in that it continues to decline as prosperity increases (Vörösmarty et al., 2010), and freshwater habitats are among the most threatened in the world (World Wildlife Fund, 2017).

A sober look at the current status of urban water infrastructure

Urban water infrastructure perpetuates an approach that attempts to fragment the water cycle. Potable water and stormwater systems are independent of each other. Wastewater is released to freshwater systems with trace contaminants that end up in drinking water due to nature's inability to assimilate these constituents, and due to the infeasibility, both technologically and financially, of wastewater treatment and water purification processes to remove them. Stormwater systems also deposit contaminants and change the amount and timing of water flowing to the natural environment. These three components of urban water infrastructure, potable water, wastewater and stormwater, were developed without due respect for the water cycle or for the aquatic environment.

The changes in our climate are manifesting to a large extent through changes to water. More intense storms, and more variable and unpredictable precipitation patterns, signal a time to reflect on past practices and future expectations for human appropriation of water resources. In addition, freshwater aquatic biodiversity is imperilled and in dire need of greater attention to its protection and rehabilitation.

Despite its inherent inefficiencies and environmental effects, the prevailing practice for urban water services is to extend centralized urban infrastructure to ever-expanding new subdivisions, because viable alternatives have not been sufficiently developed and the investments in physical plant and staff capacities favour the existing practices. The longevity and magnitude of investments in water infrastructure are such that changing directions requires a vision that extends beyond the life cycle of the current built infrastructure – well over 50 years. Even at the end of their life cycle, pipes and pumps tend to be replaced in the same configuration as the original, so the influence of new infrastructure installed today will extend many decades into the future. This time frame is politically and practically extremely challenging, in part because of election cycles that prompt political leaders to focus on time horizons of three to ten years. Political will, sustained investment, and public support are also needed to effectively manage pressures from the land development community for site plan approvals and to work within the limitations of municipal funding potentials. The reliability of the urban water infrastructure and the fact that it is largely underground, out of public sight, works against building public support and awareness of the need to shift towards more harmony with nature.

Nevertheless, our changing climate dictates that practices and design standards for urban water that are rooted in the nineteenth century cannot continue indefinitely. On the other hand, many billions of dollars' worth of urban water infrastructure cannot be changed easily or quickly. These investments, along with the associated technologies and expertise, have carved a very decisive path, which would be hard to alter for any industry, let alone one so intrinsically linked to the quality of life of communities. In addition to human health, these systems underpin a complex array of lifestyles, economic opportunities, property values, insurance rates, and health and safety standards. Millions of urban dwellers can live their entire lives without having to worry about – or even think about – water-borne disease or mass conflagration in the urban area. Along with electricity, telecommunications and roadways, water services are taken for granted in large cities with 'advanced' urban water systems.

Any plan to change course must take a long-term perspective while addressing some challenging technical issues to transition from current to future approaches. Even though all rivers flow across continents to ultimately connect with an ocean, water resources are also local. Local-scale watersheds and aquifers set the parameters for water availability, community expectations and many socio-economic conditions. For this reason, a globally defined solution to urban water resource use and infrastructure configuration is not feasible, since a one-size-fits-all solution is simply not going to work in all watersheds, aquifers and communities. However, some principles can be established for water resource use upon which a vision for future water servicing can be built. A vision could start with the question: If we were building water servicing from scratch, how would/should it be done? The ideal scenario would need to be tempered with engineering realities but will need to be envisioned by visionaries in multiple disciplines willing to think long-term and outside the conventional box. However, with a long-term vision, a plan for the realignment of urban systems can be set out and managed in an adaptive way as changes unfold.

Navigating urban waters for the future

A lot of thought has already gone into developing a set of guiding principles for urban water servicing, though an approach known as the Water Soft Path (Gleick, 2003). This approach takes as its starting point the ultimate capacity of a watershed (or aquifer) to provide water services while maintaining ecological viability. The concept reaches beyond ideas of water efficiency to question the assumed need for water in the first place (Brandes & Brooks, 2006). For instance, rather than develop demand management measures for outdoor use, a soft path approach considers the need for water services to maintain outdoor landscapes; drought-resistant plants may fulfil the need for landscaping while reducing the need for watering. The approach also matches the quality of water delivered to that needed for the end use. For instance, high-quality drinking water would not be used for toilet flushing under this approach. The Water Soft Path employs backcasting to plan from a future state, working backwards from an estimate of the sustainable yield for water extraction (Brandes & Brooks, 2006) to develop a working concept of long-term water service needs in the context of each watershed or aquifer being accessed. Conditions are also established for the quality of water returned to the ecosystem.

By starting with recognition of ecological constraints and local conditions, the Water Soft Path provides a conceptual framework for working with nature to manage water extractions in consideration of ecological needs. The basic principles include (quoted from Brooks & Brandes, 2011):

- Treating water as a service rather than an end in itself;
- Making ecological sustainability a fundamental criterion;
- Matching the quality of water delivered to that needed by the use;
- Planning from the future back to the present; and
- Ensuring community and citizen involvement in water management.

Principles for other resource use are also needed to encompass a large enough vision for urban water servicing of the future. Again, some thought has already gone into this aspect of the quandary. For instance, in 2016, the Netherlands' Parliament approved a national 'circular economy' programme, with a goal of 100% sustainable, non-polluting use of raw materials by 2050 (European Sustainable Phosphorus Platform, 2017). In a circular economy, raw materials are used and reused without harmful emissions to the environment, and new raw materials are obtained without damage to social and physical living environments in a manner that protects public health (Netherlands Ministry of Infrastructure and the Environment, Ministry of Economic Affairs, Ministry of Foreign Affairs and Ministry of the Interior and Kingdom Relations, 2016). Included in this ambitious vision is a plan to close the nutrient loop, including recovery and reuse of phosphorus and other nutrients from wastewater effluents, agricultural operations and industrial processes. The vision recognizes the importance of phosphorous as a potentially scarce resource and the environmental degradation resulting from its release in terms of eutrophication of water bodies.

While developing principles to underpin a vision is a first step, many questions remain about how urban water systems would be configured to meet these criteria. In addition, once a vision is elaborated, a plan that extends at least 100 years into the future will be needed to transition to the new configuration in communities with the conventional systems discussed above. Many of the details in these plans will need to be developed iteratively since there are some important research gaps that may change plans once results are concluded. A coordinated research agenda would assist the global community in optimizing research investments and the implementation plans for reconfigured infrastructure.

One very important and specific technical challenge concerns fire suppression. What does a future system for fire suppression look like, and what are some preliminary engineering standards? Some communities, such as Sydney, Australia (Okun, 1997), have dual pipe distribution systems that separately carry potable water and reclaimed wastewater for fire suppression. The benefits of shifting fire suppression to reclaimed wastewater sources have received some research attention (Digiano et al., 2009; Okun, 1985). Transitioning from a single potable water distribution system to a dual system is clearly cost prohibitive in the short term. However, an approach that has dual pipes installed in new subdivisions and in neighbourhoods receiving significant capital upgrades could start the transition, once the appropriate standards were established and a scheme to transmit the reclaimed wastewater developed.

Other research that could fundamentally change the established paradigm is already underway. For instance, in 2011, the Bill and Melinda Gates Foundation sponsored a challenge to create a toilet that recovers energy, water and nutrients while removing germs; operates without connection to water, sewer or electricity grids; costs less than five cents per user per day; promotes sustainable and profitable sanitation services and businesses; and is an aspirational product with potential for use everywhere (Gates Foundation, 2013).

Many additional challenges persist, including addressing trace contaminants in urban wastewaters and ensuring sewage collection systems function as the transition to lower water use is made. There are social and cultural challenges as well, and a need for sustained political will to ensure that communities are engaged and that funding is allocated to urban water infrastructure for the future.

An expanded role for smart water management

Currently, there are options for each engineered water services system to have dedicated, and in many cases advanced and elegant, automated and semi-automated management systems. Supervisory control and data acquisition systems, for example, provide essential information and support to ensure that potable water and wastewater systems operate reliably, while using energy and chemicals efficiently. This is fine as far as it goes, but the urban water complex also includes, and in fact relies on, the natural streams, creeks, rivers and lakes flowing through each urban area. These features are more than just drainage for stormwater or receiving waters for wastewater effluents. They provide one of the most important functions for successful stormwater and wastewater management in that they receive and convey the outputs of engineered urban water systems. For potable water systems, surface waters and aquifers clearly provide the most important service: water itself.

How much monitoring is done for these systems to inform managers about the potential limits of urban demands and uses? Other than flood advisories and beach water quality, what information do municipalities ensure they have on hand to best manage regional water resources in a comprehensive manner? This is not a technical issue. Smart water management techniques exist to monitor dissolved oxygen levels in surface water and to assess total suspended solids, nutrient enrichment, erosion rates and many other variables. Yet these features are not typically monitored by municipalities or water service providers because they do not consider natural systems part of the infrastructure for which they are responsible.

Just as the three components of urban water servicing can no longer be thought of as systems that are independent of each other, the natural water environment cannot be separated from the engineered systems in urban environments. When recognizing the critical role of natural drainage features for urban water services, it is equally important to recognize that these systems also provide numerous other essential services on a planetary scale. In addition to habitat for aquatic, riparian and avian species, surface water systems support nutrient cycling and carbon sequestration, and often also sustain a variety of plants and pollinators. River networks flowing through urban areas can provide migration corridors and habitat connectivity for species that would otherwise not be able to traverse urban landscapes. Green spaces along shorelines provide human populations with numerous opportunities to maintain their physical and mental health. We need to pay attention to the altered flows, water chemistry and temperatures of urban water environments to reverse the dire trends of globally imperilled freshwater species.

Hence, when developing the principles for water systems of the future, the natural aquatic environment cannot be excluded, not only for its service to urban infrastructure but also for its invaluable ecological services. The smart water management tools are at hand; what is lacking is recognition of the need for careful monitoring and nurturing of urban water networks.

Conclusion

Urban water infrastructure has evolved over the past century without taking a required pause to assess whether carrying forward the logic of the past makes sense for a future with greater urban population density, highly stressed aquatic ecosystems and a changing climate. Although it is vital to apply smart water techniques to optimize the existing infrastructure, since we will be relying on it for many years to come, it is equally important to develop a long-term vision for urban water infrastructure that respects the water cycle and the need for all life forms to have access to water and suitable habitat. The water cycle will prevail no matter what engineered systems are in place. Water cannot be segmented for the purposes of various human needs since it operates in a cycle and it is intimately connected with the land on which it falls, over which it runs and into which it percolates. It is also home to many species, the preservation of which is necessary to fulfil many human needs but also for their intrinsic value and beauty.

As a first step, a set of guiding principles for water servicing could be developed through consultations with leaders, communities and experts. These principles could be used to develop a vision for urban water infrastructure of the future that can be adapted to local watershed and aquifer conditions. Even though individual local conditions will require tailor-made plans, there are many common challenges, and a coordinated research agenda could support implementation of the plans and the vision.

> *The health of land and water – and of woods, which are the keepers of water – can be the only lasting basis for any civilization's survival and success.*
>
> —R. Wright, *A Short History of Progress* (2004)

Disclosure statement

No potential conflict of interest was reported by the author.

ORCID

M. P. Trudeau http://orcid.org/0000-0001-8603-9771

References

Adamowski, J., & Prokoph, A. (2013). Assessing the impacts of the urban heat island effect on streamflow patterns in Ottawa, Canada. *Journal of Hydrology*, *496*(1), 225–237. https://doi.org/10.1016/j.jhydrol.2013.05.032

American Society of Civil Engineers. (2016). *Failure to act: Closing the infrastructure investment gap for America's economic future*. Reston, VA. Retrieved February 2020 from https://www.infrastructurereportcard.org/wp-content/uploads/2016/10/ASCE-Failure-to-Act-2016-FINAL.pdf

Benidickson, J. (2007). *The culture of flushing: A social and legal history of sewage*. UBC Press.

Bipartisan Policy Center. (2017). *Understanding America's water and wastewater challenges*.

Boers, T., DeGraaf, M., Feddes, R. A., & Ben-Asher, J. (1986). A linear regression model combined with a soil water balance model to design micro-catchment for water harvesting in arid zones. *Agricultural Water Management*, *11*(3–4), 187–206. https://doi.org/10.1016/0378-3774(86)90038-7

Brabec, E., Schulte, S., & Richards, P. (2002). Impervious surfaces and water quality: A review of current literature and its implications for watershed planning. *Journal of Planning Literature*, *16*(4), 499–514. https://doi.org/10.1177/088541202400903563

Brandes, O., & Brooks, D. (2006). *The soft path for water in a nutshell*. Friends of the Earth Canada and the POLIS Project on Ecological Governance. University of Victoria.

Brooks, D., & Brandes, O. M. (2011). Why a water soft path, why now and what then? *Water Resources Development*, *27*(2), 315–344. https://doi.org/10.1080/07900627.2011.571235

Canadian Infrastructure Report Card. (2019). *Monitoring the state of Canada's core public infrastructure*. Retrieved February 2020 from http://canadianinfrastructure.ca/downloads/canadian-infrastructure-report-card-2019.pdf

City of Ottawa. (2001). *Environmental Services Committee minutes*. Retrieved February 2020 from https://app06.ottawa.ca/calendar/ottawa/citycouncil/esc/2001/05-22/minutes.htm

City of Ottawa. (2013). *Infrastructure master plan, Chapter 5.0: Water, wastewater, stormwater and village infrastructure plans*. Retrieved February 2020 from https://documents.ottawa.ca/sites/documents/files/documents/imp_chapter_5_1_5_3_en.pdf

City of Toronto. (2009). *Design criteria for sewers and watermains* (1st ed.). Technical Services.

Collivignarelli, M., Abbà, A., Frattarola, A., Carnevale Miino, M., Padovani, S., Katsoyiannis, I., & Torretta, V. (2019). Legislation for the reuse of biosolids on agricultural land in Europe: Overview. *Sustainability*, *11*(6015), 22. https://doi.org/10.3390/su11216015

Costa, M. H. A., Botta, A., & Cardille, J. A. (2003). Effects of large-scale changes in land cover on the discharge of the Tocantins River, Southeastern Amazonia. *Journal of Hydrology*, *283*(1–4), 206–217. https://doi.org/10.1016/S0022-1694(03)00267-1

CVCA and TRCA. (2010). *Low impact development stormwater management planning and design guide, version 1.0*. Credit Valley Conservation Authority [CVCA] and Toronto Region Conservation Authority, Downsview and Mississauga.

Deb, A. K., & Evans, M. P. (1980). Dual distribution systems analysis. *Journal – American Water Works Association*, *72*, 103–108. https://doi.org/10.1002/j.1551-8833.1980.tb04475.x

Digiano, F. A., Weaver, C. C., & Okun, D. A. (2009, February). Benefits of shifting fire protection to reclaimed water. *American Water Works Association Journal*, *101*(2), Business Premium Collection pg. 65. https://doi.org/10.1002/j.1551-8833.2009.tb09840.x

Dudgeon, D., Arthington, A. H., Gessner, M. O., Kawabata, Z., Knowler, D. J., Lévêque, C., Naiman, R. J., Prieur-Richard, A., Soto, D., Stiassny, M. L., & Sullivan, C. A. (2006). Freshwater biodiversity: Importance, threats, status and conservation challenges. *Biological Reviews of the Cambridge Philosophical Society*, *81*(2), 163–182. https://doi.org/10.1017/S1464793105006950

European Sustainable Phosphorus Platform. (2017). *ESPP eNews no. 6, January 2017*. Retrieved February 2020 from http://www.phosphorusplatform.eu/scope-in-print/news/1398-enews6#_Toc472347497

Gates Foundation. (2013). *Water, sanitation and hygiene: Reinvent the Toilet Challenge fact sheet*. Retrieved February 2020 from https://docs.gatesfoundation.org/Documents/Fact_Sheet_Reinvent_the_Toilet_Challenge.pdf

Gleick, P. H. (2003). Global freshwater resources: The concept of water soft paths. *Science*, *302* (5650), 1524–1528. https://doi.org/10.1126/science.1089967

Hamid, H., & Eskicioglu, C. (2012). Fate of estrogenic hormones in wastewater and sludge treatment: A review of properties and analytical detection techniques in sludge matrix. *Water Research*, *46*(18), 5813–5833. https://doi.org/10.1016/j.watres.2012.08.002

Hickey, H. (2008). *Water supply systems and evaluation methods, Volume 1: Water supply system concepts*. US Fire Administration, FEMA, US Department of Homeland Security. Retrieved February 2020 from https://www.usfa.fema.gov/downloads/pdf/publications/water_supply_systems_volume_i.pdf

Höglund, C. (2001). *Evaluation of microbial health risks associated with the reuse of source separated human urine* [PhD thesis]. Department of Biotechnology, Royal Institute of Technology. ISBN 91-7283-039-5.

Kidd, K. A., Blanchfield, P. J., Mills, K. H., Palace, V. P., Evans, R. E., Lazorchak, J. M., & Flick, R. W. (2007). Collapse of a fish population after exposure to a synthetic estrogen. *Proceedings of the National Academy of Sciences*, *104*(21), 8897–8901. https://doi.org/10.1073/pnas.0609568104

Klein, R. D. (1979). Urbanization and stream quality impairment. *Journal of the American Water Resources Association*, *15*(4), 948–963. https://doi.org/10.1111/j.1752-1688.1979.tb01074.x

Löfvenhaft, K., Runborg, S., & Sjögren-Gulve, P. (2004). Biotope patterns and amphibian distribution as assessment tools in urban landscape planning. *Landscape and Urban Planning*, *68*(4), 403–427. https://doi.org/10.1016/S0169-2046(03)00154-3

Marcogliese, D. J., Blaise, C., Cyr, D., de Lafontaine, Y., Fournier, M., Gagné, F., Gagnon, C., & Hudon, C. (2015). Effects of a major municipal effluent on the St. Lawrence River: A case study. *Ambio*, *44*(4), 257–274. https://doi.org/10.1007/s13280-014-0577-9

Netherlands Ministry of Infrastructure and the Environment, Ministry of Economic Affairs, Ministry of Foreign Affairs and Ministry of the Interior and Kingdom Relations. (2016). *A circular economy in the Netherlands by 2050*. Retrieved February 2020 from https://www.government.nl/documents/policy-notes/2016/09/14/a-circular-economy-in-the-netherlands-by-2050

Okun, D. A. (1985). Reuse: Panacea or pie in the sky? *American Water Works Association Journal*, *77*(7), Water reuse: The second time around (July 1985), p. 26. https://doi.org/10.1002/j.1551-8833.1985.tb05562.x

Okun, D. A. (1997). Distributing reclaimed water through dual systems. *American Water Works Association Journal*, *89*(11), 52. https://doi.org/10.1002/j.1551-8833.1997.tb08321.x

Paul, M. J., & Meyer, J. L. (2001). Streams in the urban landscape. *Annual Review of Ecology and Systematics*, *32*(1), 333–365. https://doi.org/10.1146/annurev.ecolsys.32.081501.114040

Powell, J. (2014). *The Great Fire of 1870, today in Ottawa's history*. Retrieved February 2020 from https://todayinottawashistory.wordpress.com/2014/08/20/the-great-fire-of-1870/

Schindler, D. W., Hecky, R. E., & McCullough, G. K. (2012). The rapid eutrophication of Lake Winnipeg: Greening under global change. *Journal of Great Lakes Research*, *38*(Supplement 3), 6–13. https://doi.org/10.1016/j.jglr.2012.04.003

Schueler, T. (1994). The importance of imperviousness. *Watershed Protection Techniques*, *1*(3), 100–111.

Shen, L., Fischer, J., Martin, J., Hoque, M. E., Telgmann, L., Hintelmann, H., Metcalfe, C. D., & Yargeau, V. (2016). Carbon Nanotube Integrative Sampler (CNIS) for passive sampling of nanosilver in the aquatic environment. *Science of the Total Environment*, *569–570*(Complete), 223–233. https://doi.org/10.1016/j.scitotenv.2016.06.095

Stanfield, L., Gibson, S., & Borwick, J. (2006). Using a landscape approach to identify the distribution and density patterns of salmonids in Lake Ontario Tributaries. *American Fisheries Society Symposium*, *48*, 601–621.

Stanfield, L., & Kilgour, B. (2006). Effects of percent impervious cover on fish and benthos assemblages and instream habitats in Lake Ontario tributaries. *American Fisheries Society Symposium*, *48*, 577–599.

Swedish International Development Cooperation Agency & Esrey, S. (Ed). (2001). *Closing the loop: Ecological sanitation for food security*. Publications on Water Resources No. 18, UNDP and Sida. Retrieved February 2020 from http://www.ecosanres.org/pdf_files/closing-the-loop.pdf

Trudeau, M. P., & Morin, A. (2017). Associations of event-scale flow hydrology with fish richness in urbanizing Canadian watersheds of Lake Ontario. *Ecohydrology*, *10*(3), e1087. https://doi.org/10.1002/eco.1807

Trudeau, M. P., & Richardson, M. (2016). Empirical assessment of watershed scale effects of urbanization on event flow hydrology in watersheds of Canada's Great Lakes - St. Lawrence Basin. *Journal of Hydrology*, *541 B*, 1456–1474. https://doi.org/10.1016/j.jhydrol.2016.08.051

Vörösmarty, C. J., McIntyre, P. B., Gessner, M. O., Dudgeon, D., Prusevich, A., Green, P., Glidden, S., Bunn, S. E., Sullivan, C. A., Liermann, C. R., & Davies, P. M. (2010). Global threats to human water security and river biodiversity. *Nature*, *467*(7315), 555–561. https://doi.org/10.1038/nature09440

Walsh, C. J., Roy, A. H., Feminella, J. W., Cottingham, P. D., Groffman, P. M., & Morgan, R. P. (2005). The urban stream syndrome: Current knowledge and the search for a cure. *Journal of the North American Benthological Society*, *24*(3), 706–723. https://doi.org/10.1899/04-028.1

Weber, T. (2018). *What you need to know about ISO's 2018 public protection classification*. FireRescue 1. Retrieved February, 2020 from https://www.firerescue1.com/community/articles/what-you-need-to-know-about-isos-2018-public-protection-classification-VqnTbt16kpon82d2/

World Health Organization. (2017). *Guidelines for drinking water quality* (4th ed.). Retrieved February 2020 from https://www.who.int/water_sanitation_health/publications/drinking-water-quality-guidelines-4-including-1st-addendum/en/

World Wildlife Fund. (2017). *Freshwater habitat: Overview*. Retrieved February 2020 from https://www.worldwildlife.org/habitats/freshwaters

Wright, R. (2004). *A short history of progress*. Anansi.

The moral hazards of smart water management

Kris Hartley and Glen Kuecker

ABSTRACT

Smart water management (SWM) brings technological sophistication to water governance by providing monitoring, operational and communications capacities through real-time information. SWM's quantification appeals to metric-driven governance but, we argue, also perpetuates a technocratic and instrumental-rationalist mindset. The peril of this mindset is that it sees technology as a solution for sustainability problems caused by deep-seated structural and behavioural faults. This essay reflects on this dynamic by siting the SWM concept within discussions about technocracy, moral hazard and power dynamics. It suggests that SWM's rhetorical positioning undermines its own goals while naively seeking universal applicability, resolvable by embracing the precautionary principle.

Introduction: reflecting on smart water management

The increasing sophistication and proliferation of 'smart' technologies for urban, resource, and infrastructure governance not only mark a sunrise moment for smart water management (SWM) but also herald a deeper entrenchment of technocracy's influence on the narratives and practices of the global sustainability project. The influence of SWM is likely to grow as governments implement its principles and declare success based on selected metrics. It is appropriate, however, not to lose a critical theoretical perspective that identifies and challenges technocratic power within SWM. By 'critical,' we do not mean a metric-based evaluation of SWM but an effort to understand the manifestations and exercises of elite interests within SWM. This essay explores these dynamics from the perspective of technocracy and instrumental rationalism and offers recommendations for how SWM can avoid perpetuating a sustainability narrative that compromises its own stated goals.

Smart water management (SWM) is the use of technology-based systems to provide real-time information for efficient, effective, and collaborative or automated water management. Policy actors, after years of refining it through digital transformation, increasingly embrace SWM as a tool for facilitating integrated water resources management (IWRM) across manifold policy domains including sourcing, delivery, quality, and resilience amidst external threats like floods and droughts. Policy actors see SWM also as a pathway towards broader social, economic, environmental, and governance objectives, as consistent with the visions of the United Nations Sustainable Development Goals (SDGs). South Korea is a leader in the development and rollout of SWM and proponents of SWM are promulgating the country's experience as an example for other countries.

The maturity of the SWM concept and the enthusiasm with which policy actors now embrace it signals an appropriate time for abstract reflection about the political dimensions of SWM. The resulting insights can give greater legitimizing effect to SWM by helping address its limitations. Our provocation is that SWM risks perpetuating a technocratic mindset that is blinded to the origins of the sustainability crisis. We contend that SWM is at the trailhead of the following path: (i) technology buys society more time to luxuriate in convenient and indulgent consumption habits; (ii) technology thereby excuses society from a painful reckoning about the origins and perils of these habits; and (iii) society anticipates that a continuous flow of novel technologies will always offset the negative effects of increasing unsustainability – even as populations grow and lifestyle-based ecological footprints expand.

The actionable corollary is that SWM's underlying logic of efficient optimization undermines its own broader pursuit of sustainability, with remediation possible either through a synthesis of 'hard' (purely mechanical) and 'soft' (behavioural and conceptual) approaches or through a recognition of the limits and hazards of hard approaches and an effort to apply them to only the most technical problems. Either would seek to avoid cooptation of soft approaches by the logical dominance and popularity of hard approaches. In seeking to understand the process by which SWM undermines the conceptualization and pursuit of sustainability, and in proposing a way to address this dilemma, this essay examines the logic of technocratic policymaking, applies it to SWM and the global dissemination of its narrative, and considers how society might reimagine technocracy. Rather than being a purely iconoclastic exercise, this essay aims to inspire policymakers to think reflectively about SWM and their own role in its promotion.

This essay continues by examining the epistemic roots of technocracy and instrumental rationalism and their contribution to moral hazard. Thereafter, we consider how technocratic reasoning in SWM presents itself as a solution to the moral hazard it creates while ignoring the reproduction of power dynamics and underlying determinants of the sustainability crisis. Finally, we interrogate the narrative hegemony and consequent global transferability of SWM and technocracy more generally, critically re-centring the concept without categorically dismissing its objectives. We close with an appeal to the precautionary principle, calling for more disciplined reflection, intensive inquiry, and good-faith cross-examination within a policy epistemic otherwise characterized by exclusionary self-certitude.

Instrumental rationalism and the logic of technocratic policymaking

Technocratic approaches are often mismatched with complex or 'wicked' policy problems (Hartley et al., 2019; Head, 2019). The narrative perpetuation of a nature-culture split[1] reproduces the technocratic epistemic by promoting instrumental-rationalist logic[2] in the management of natural and human realms that are resistant to reductive governance. Taking a technocratic approach to SWM, for example, could lead to an entrenched efficiency trap that profoundly exacerbates water sustainability challenges.[3] The mechanism connecting the efficiency trap and sustainability crisis is moral hazard; in this case, how society's consumption behaviours assume and come to rely on technical fixes for material challenges arising from declining sustainability. The term 'moral hazard,' as we use it, does not invoke the colloquial definition of 'moral' as 'ethical,' but instead refers to the behavioural economic concept in a principal-agent setting characterized by imperfect information. The concept is commonly illustrated in terms of insurance. Marshall (1976) provides the following definition: 'moral hazard ... is defined to be any misallocation of resources which results when risks are insured with normal insurance contracts and only with such contracts. In this case "normal" means that the insured is paid a specified amount in case of loss but his actions in avoiding or reducing loss are not stipulated by the contract' (p. 880).

The analogue of moral hazard to the sustainability crisis asks for no great leap of logic. Technology is 'insurance' against actions of society that inadequately avoid or reduce loss from ecological overshoot and its consequences (Kuecker & Hall, 2011; Tainter, 1988). In short, society and individuals avoid behavioural change because they anticipate being 'bailed out' by increasingly sophisticated technologies. Furthermore, the use of moral hazard to describe SWM's potential challenges highlights the externalization or socialization of costs, which in this case are not only the real costs of infrastructure development but also the abstract and real costs of declining sustainability resulting from the continued or increased consumption needed to maintain consumption habits. This perspective shifts the conceptualization of water from a 'toll good' to a 'common pool resource,' changing the relationship between society and nature to one of conservation and survival rather than of extraction and contract-based market exchange (Hartley, 2018).

Instrumental rationalism, the applied manifestation of technocracy partially responsible for exacerbating moral hazard, is the proverbial hammer that treats every problem like a nail; the broken pipe is fixed without deeper reflection about why it was broken. Behaviours that broke the pipe fail to change because the fix is always available, invoking moral hazard by illustrating the 'safety' of maintaining existing habits and the externalization or socialization of costs associated with them. In service to short-term fiscal convenience and political feasibility, the reductionist 'problem-solving' or 'management' epistemic overlooks predicaments and wickedness for which conclusive solutions are inconveniently unavailable. Despite these shortcomings, the technocratic epistemic (referring more to an idea than to the people or interests executing it) maintains its practical legitimacy because it boasts a record of delivering immediately tangible solutions and performing well on a curated suite of myopic fiscal and operational measures.

By operating in this way, the epistemic is never politically pressured to confront the structural determinants of wicked policy problems and the embeddedness of those problems in the technocratic solutions proposed for them (e.g., technological initiatives as an outlet for surplus investment capital – evidence of the dominant market-based and consumption-based logics testing ecological limits). While the 'common-sense' narrative of technocratic solutions, like various iterations of 'smart,' enjoys favour among political and corporate elites, the current era of populist agitation has shown that such dominant narratives are assailable; indeed, Caesar's wife is not above suspicion. Where technocratic solutions contravene populist sentiment, they lose their legitimacy amidst politically charged rhetoric. This was evident when United States President Donald Trump undermined the authority and popular credibility of scientific experts, including the Centres for Disease Control (CDC), in seeking to prematurely 're-open' the economy amidst a continuing increase in COVID-19 cases.[4] This case underscores that claims to legitimacy made by technocrats are not impervious to political challenge, a matter that policymakers must consider when making appeals to the authority of expertise.

Finally, while technocratic thinking exists in ring-fenced policy domains like SWM and smart cities, global institutional architecture like that supporting the SDGs is a type of Trojan Horse through which such ideas are widely disseminated (e.g., concepts like 'best practices' and 'first principles' that *prima facie* do not invite criticism but are subtly deployed to privilege certain policy narratives). The normalizing power of the instrumental-rationalist perspective is given effect by an enabling global political economy of development, reified through actors, ideas, and institutions (Béland, 2009): actors and their relationships as constitutive of a power-knowledge nexus, ideas as the narratives around sustainability used for validating public resource commitments, and 'hard' institutions that protect and facilitate those ideas – including local or national governance arrangements (e.g., special-purpose investment vehicles) on one hand and the SDGs and associated institutional structures on the other. In this way, 'smart' practices like SWM progress towards 'common-sense' status, promulgated by policy actors as exemplars of successful experimentation. This process gives rise to a core-periphery dynamic in the translation and transfer of hegemonic policy ideas, resulting in a disciplined and normalized view of problems and their solutions; logical shortcomings are widely disseminated but weakly interrogated.

Smart water management as technocratic thinking

In smart initiatives like SWM, the private sector produces and mainstreams science- and technology-backed initiatives around which the logic of naming and framing problems is designed to fit the solutions on-offer.[5] This process illustrates how capitalist logic induces the types of investments that perpetuate particular approaches to policy intervention. The accompanying narratives around pragmatism and the primacy of technology support the political legitimacy of such interventions and congeal over time to form the 'common-sense' logic behind buzz-solutions like smart cities and 'disaster risk reduction.' As the conceptual impetus behind a revised (but not uprooted) narrative drawing on longstanding capitalist doctrine, 'smart' or 'sustainable' become a new 'development' (as a globally encompassing policy vision) through which technology provides solutions and markets drive needed innovation and mainstreaming (Kuecker & Hartley, 2020a).

In accordance with this model, SWM is likely to be the dominant water management paradigm in service to 'new development,' due principally to its appealing allure as a scientific, evidence-based, 'common-sense' approach to IWRM and to the copiously articulated compatibility of its narrative with that of the SDGs. Both facilitate SWM's transferability in the international context, as SWM's ideas become principles through institutional privileging and the self-referential legitimacy and credibility of its expert defenders. For example, capitalist reproduction is a moral hazard that SWM obscures by discursively positioning itself as a 'sustainability-focused' solution; this itself becomes a moral hazard by externalizing the social and environmental costs of capitalist reproduction.

With SWM serving as a prescriptive and formulaic means for accomplishing the SDGs, its technocratic gestures not only generate the aforementioned moral hazard but also risk marginalizing alternative narratives about society's declining ability to sustain itself, including those that challenge capitalism's logic of perpetual growth and market fundamentalism. Nearly one billion informal dwellers around the world struggle every day to procure water, a circumstance exacerbated by the COVID-19 pandemic. The resulting incomplete or staged 'soft collapse' (Kuecker, 2020) invites questions not about the design and calibration of policy instruments – even the most technologically advanced – but about the deference that policy narratives show to a totalizing capitalist project progressing unchallenged despite imminent depletion of natural resources and a socially and politically destabilizing wealth gap. The moral hazard of SWM is that it pretends to solve or trick the laws of thermodynamics, especially the wicked problem of systemic entropy, with its ring-fenced problem-solving epistemic. The SWM solution adds additional order to a system already experiencing an unsustainable rate of declining returns from successive technocratic solutions and waves of technological advancement. Chasing the failure of order is, inevitably, more order – accelerating collapse. In short, society is doing the wrong things but more efficiently, which in a profound paradox can be politically sold as progress. Ultimately, however, an efficiency trap leads to an entropic trap, as illustrated by the Tainter curve (Tainter, 1988).[6] We contend that it is time to break this cycle.

The pursuit of more order will not end without a painful reckoning. Like its recent smart city adventures (G. D. Kuecker & Hartley, 2020b), South Korea's SWM efforts are backed by world-class technology, ample capital, and a competent government committed to its own vision for the future. SWM will be highly successful according to its principal metrics: efficiency and effectiveness in service delivery, as against broader reforms to economic and societal systems that reduce the need for such urgent and aggressive technocratic policy action. Further, we maintain that the leaders, technicians, and scholars driving South Korea's SWM project are guided by a genuine interest in addressing water challenges and a sincere belief in the efficacy of their methods. As such, this critique is based on the overarching and hegemonic ideas that have come to dominate the sustainability narrative worldwide and reinforce the influence of the power-knowledge nexus. The executors of SWM may fail to see, for example, the moral hazard of technocratic interventions because the policy epistemic by whose language problems are named and framed dominates the way these agents are trained and 'socialized' into the community of technocratic practice. The failing is not reflective of the actions and motivations of individuals within the system but of macro-structural forces that are difficult to see and beyond any individual's influence. In this way, the question of SWM legitimacy illustrates the predicaments of cultural hegemony as well as it does the Marxist question of structure and agency as explored in the context of rational choice (Carver & Thomas, 2016; Carling, 1995) and collective action (Brewer, 1987; Lash & Urry, 1984).

While smart technology promises to bring sustainable efficiencies to the challenges of water management, they present an additional, longer-term moral hazard of unintended consequences. Lurking within SWM's wires, monitors, routers, and clouds are powerful algorithms that pursue technocratic rationalism's dream of perfecting the human condition. The promise of these algorithms, especially with the emerging power of quantum computing, offer society the ultimate insurance policy against the perils of ignoring existential threats to the planetary system. With this algorithm comes a profound moment of transition, the emergence of 'singularity' (Shanahan, 2015; Vinge, 1993) that accomplishes the ultimate act of problem-solving by removing human agency from the pursuit of sustainability. SWM's moral hazard faces an imminent reckoning about this post-human Pandora's box, one that cannot simply be dismissed until it fully reveals itself.

Towards even 'smarter' water management

There is no easy way to question the received technocratic narrative in public policy because the mechanisms by which it interprets the world have no way of recognizing alternatives; examples are the concepts of governance and public policy itself (neither of which have a language to engage meaningfully with the concept of anarchy, for example) and the concepts of capitalism and markets (which have no language to describe alternative means of resource distribution). Alternative ideas are not so much confronted and discredited by technocracy as simply erased or rendered discursively invisible by its epistemic frame, and there is no critically self-reflective tool that allows the technocratic epistemic to stand outside of itself and ask prickly questions. As a programmed or algorithmic concept, technocracy sees only what it is told to measure and manages only what it sees. Any self-vetting would be done only within the confines of the same epistemic that created the policy to begin with – the core flaw in an SDG logic that so eagerly presents itself as a transformative and emancipatory way of thinking. Even the concept of community empowerment, which ostensibly plays a role in SWM and receives copious mention in related policy documents,[7] exists at the pleasure of the epistemic that defines the macro-concept of SWM (i.e., that 'smart' as technical and 'management' as technocratic or capitalistic are pathways to sustainability, with engagement efforts asking only which flavour of technocracy and capitalism the public prefers). It is often in the halls of academia where critical vetting takes place, illustrating the importance of scholarly independence and, as importantly, the perilous degree to which technocratic and instrumental-rationalist epistemics have captured governance institutions.

One way to address the hegemony of technocracy and SWM's consequent moral hazard is to apply the precautionary principle to its planning, design, and implementation. The precautionary principle suggests a range of metrics around limiting potential harm by regulating or barring policies especially when the prospects of harm are unknown. It restrains the Cartesian proclivities of instrumental rationalism, which presume that humans can discipline nature through reason; in this way, the principle implicitly and explicitly recognizes the rights of nature while also presuming the fallibility of human reason. This approach invites policymakers to assume that if something can go wrong with SWM, then it will – a logic that commands them to consider all regulatory contingencies, including cancelation. Sunstein (2005) calls the precautionary principle 'the laws of fear' that lead inevitably to policy paralysis, a critique that resonates in a world currently facing a 'perfect storm' of large-scale, intersecting, and synchronous crises, among them water scarcity (Kuecker, 2014). At the crux of the predicament of balancing policy urgency with mitigating moral hazards is the question of risk, especially in its socio-cultural manifestation. The conceptualization of 'risk society' (Beck et al., 1992; Jarvis, 2007), as a fundamental feature of the modern policy epistemic in how it structures markets and managerialist cost-benefit analysis, is constitutive of what earns technologies their 'smart' distinction; technocrats identify, parameterize, and eliminate risk. However, the perfect storm, now potentially manifest as a pandemic-driven soft collapse (Kuecker, 2020), destabilizes the structures that constitute the meaning of risk itself, to the degree that the constitutive position of risk in defining the need for 'insurance' falters in a world of multiplying crises and moral hazards. Technocracy, as the epistemic underwriting risk and SWM, becomes a dysfunctional construct that is as eligible for policy exile as a blithely ignored set of CDC social distancing guidelines.

Politically risky though it may appear, public policy would benefit from getting comfortable with a critical view of its own epistemics – which for practical reasons should be expressed in friendly critiques that provoke but do not alienate. The question then becomes how policy makers and 'street-level bureaucrats' (Lipsky, 2010) can stand watch against regressive and costly reproductions of power-knowledge narratives like 'smart,' 'development,' and 'sustainability' – terms laden with normative ideas and agendas but enjoying canonical and virtually unassailable status. While our recommended solution is to deploy mechanisms for vetting these and other moral hazards, the difficulty is that such undertakings are done often as mere institutional window-dressing (e.g., by appointing a committee or adding a paragraph in a consulting report) while actually requiring a fundamental change in mindset – not from the 'wrong' old thing to the 'right' new thing, but from stable and occasionally arrogant 'certitude' to precaution,[8] disciplined reflection, intensive inquiry, and good-faith cross-examination. This solution requires the type of sincerely self-critical attitude that leadership in many countries – particularly those under the power of dominant political parties desperate to maintain their legitimacy – simply cannot have. However, in freer multi-party democracies like South Korea, Taiwan, Japan, Australia, the UK, Germany, and others, such a change in mindset is a possibility.

South Korea's SWM project will likely earn globally elite 'pilot' status, and its narrative convergence with the SDG project will further the privileging of its instrumental-rationalist elements in global narratives about the 21st century's existential challenges. In closing, however, our critical argument is made not in a fit of grizzled pessimism but in the sincere hope that the same creativity, ambition, and competence that South Korea brings to innovative megaprojects like SWM and smart cities will inspire policy actors to undertake a fundamental re-thinking of technocratic governance and ultimately to critically reflect on the hazardous logic on which modern society is precariously situated. On this issue, South Korea has an opportunity to be a transformative global leader.

Notes

1. Illustrating the concept of the nature-culture split, Kuecker and Hartley (2020a) argue that 'the societal impacts of natural disasters are socially constructed by [for example, the disaster risk reduction narrative] as insults to reason – a rebellion by nature – rather than a consequence of modernity's faulty attempt to rationalize nature' (p. 10).
2. This essay uses the term 'rationalism' rather than 'rationality.' The terms may appear to be interchangeable but there is a subtle and important distinction. Rationality is the behavioural logic that observes a particular rule-set (e.g., the self-interested behaviour of an economic 'satisficer'). Rationalism connotes a type of signalling in which individual or institutional behaviour (e.g., policymaking) embraces 'rationality' for specifically political, social, economic, or organizational objectives. Referring to rationalism primarily as the 'rule

of reason,' we often preface the term with the qualifier 'instrumental' (i.e., the tools of policy) or 'technocratic' (e.g., the epistemic embraced by policymaking systems).

3. The efficiency trap (Hallett, 2013) has long been theorized through the 'Jevons paradox' (Alcott, 2005), in which the more efficient extraction, processing, and/or use of a resource is accompanied by increased demand for and use of that resource. While examples of the Jevons paradox have been used commonly in reference to natural resources, the study of transportation planning has also applied it to the concept of 'induced demand,' in which the expansion of transportation capacity (e.g., addition of new lanes on a highway) leads to increased usage that offsets the efficiency gains of the added capacity (Cervero, 2002). The application to SWM and the relationship between water supply and demand adopts the same logic.
4. https://www.nytimes.com/2020/05/07/us/politics/trump-cdc.html (accessed 9 July 2020)
5. This rhetorical sleight-of-hand is reminiscent of the real estate agent telling a client, 'if you bought this house, you would be home by now.' The problem (that the client is, at that moment, not currently at home) is constructed in a way that can solved by the item already for sale.
6. The Tainter curve (Tainter, 1988) is an n-shaped graphic showing that the benefits of complexity within a system initially rise as the level of complexity increases, but at a declining rate. Past a given inflection point, further increasing levels of complexity coincide with a decline in the benefits of complexity.
7. https://www.iwra.org/swmreport/(accessed 9 July 2020)
8. According to O'Riordan and Jordan (1995), 'At the core of the precautionary principle is the intuitively simple idea that decision makers should act in advance of scientific certainty to protect the environment (and with it the well being interests of future generations) from incurring harm ... In essence, it requires that risk avoidance becomes an established decision norm where there is reasonable uncertainty regarding possible environmental damage or social deprivation arising out of a proposed course of action' (p. 3).

Disclosure statement

No potential conflict of interest was reported by the authors.

ORCID

Kris Hartley http://orcid.org/0000-0001-5349-0427
Glen Kuecker http://orcid.org/0000-0001-5522-2821

References

Alcott, B. (2005). Jevons' paradox. *Ecological Economics*, *54*(1), 9–21. https://doi.org/10.1016/j.ecolecon.2005.03.020

Beck, U., Lash, S., & Wynne, B. (1992). *Risk society: Towards a new modernity* (Vol. 17). Sage Press.

Béland, D. (2009). Ideas, institutions, and policy change. *Journal of European Public Policy*, *16*(5), 701–718. https://doi.org/10.1080/13501760902983382

Brewer, J. (1987). Exploitation in the new Marxism of collective action. *The Sociological Review*, *35* (1), 84–96. https://doi.org/10.1111/j.1467-954X.1987.tb00004.x

Carling, A. (1995). Rational choice marxism. In T. Carver & P. Thomas (Eds.), *Rational choice marxism* (pp. 31–78). Palgrave Macmillan.

Carver, T., & Thomas, P. (eds.). (2016). *Rational choice Marxism*. Springer.

Cervero, R. (2002). Induced travel demand: Research design, empirical evidence, and normative policies. *Journal of Planning Literature*, *17*(1), 3–20. https://doi.org/10.1177/088122017001001

Hallett, S. (2013). *The efficiency trap finding a better way to achieve a sustainable energy future*. Prometheus Books.
Hartley, K. (2018). Environmental resilience and intergovernmental collaboration in the Pearl River Delta. *International Journal of Water Resources Development*, *34*(4), 525–546. https://doi.org/10.1080/07900627.2017.1382334
Hartley, K., Kuecker, G., & Woo, J. J. (2019). Practicing public policy in an age of disruption. *Policy Design and Practice*, *2*(2), 163–181. https://doi.org/10.1080/25741292.2019.1622276
Head, B. W. (2019). Forty years of wicked problems literature: Forging closer links to policy studies. *Policy and Society*, *38*(2), 180–197. https://doi.org/10.1080/14494035.2018.1488797
Jarvis, D. S. (2007). Risk, globalisation and the state: A critical appraisal of Ulrich Beck and the world risk society thesis. *Global Society*, *21*(1), 23–46. https://doi.org/10.1080/13600820601116468
Kuecker, G. D. (2014). The perfect storm: Catastrophic collapse in the 21st century. In D. Humphreys & S. Stober (Eds.), *Transitions to sustainability: Theoretical debates for a changing planet* (pp. 89–105). Common Ground Publishing.
Kuecker, G. D. (2020). The perfect storm's pandemic driven soft collapse. *The International Journal of Environmental, Cultural, Economic and Social Sustainability*. Forthcoming.
Kuecker, G. D., & Hartley, K. (2020a). Disaster risk reduction and the development narrative: Towards a new public policy epistemic. In A. Brik & L. Pal (Eds.), *The future of the policy sciences*. Edward Elgar Publishing. Forthcoming.
Kuecker, G. D., & Hall, T. D. (2011). Resilience and community in the age of world-system collapse. *Nature and Culture*, *6*(1), 18–40. https://doi.org/10.3167/nc.2011.060102
Kuecker, G. D., & Hartley, K. (2020b). How smart cities became the urban norm: Power and knowledge in New Songdo City. *Annals of the American Association of Geographers*, *110*(2), 516–524. https://doi.org/10.1080/24694452.2019.1617102
Lash, S., & Urry, J. (1984). The new Marxism of collective action: A critical analysis. *Sociology*, *18* (1), 33–50. https://doi.org/10.1177/0038038584018001004
Lipsky, M. (2010). *Street-level bureaucracy: Dilemmas of the individual in public service*. Russell Sage Foundation.
Marshall, J. M. (1976). Moral hazard. *The American Economic Review*, *66*(5), 880–890.
O'Riordan, T., & Jordan, A. (1995). The precautionary principle, science, politics and ethics. *CSERGE Working Paper PA 95-02*. Centre for Social and Economic Research on the Global Environment.
Shanahan, M. (2015). *The technological singularity*. MIT press.
Sunstein, C. R. (2005). *Laws of fear: Beyond the precautionary principle* (Vol. 6). Cambridge University Press.
Tainter, J. (1988). *The collapse of complex societies*. Cambridge University Press.
Vinge, V. (1993, March). Technological singularity. In *VISION-21 Symposium sponsored by NASA Lewis Research Center and the Ohio Aerospace Institute* (pp. 30–31). Cleveland, OH: NASA Lewis Research Center. http://www8.cs.umu.se/kurser/5DV084/HT10/utdelat/vinge.pdf

Masculinity and smart water management: why we need a critical perspective

Anna Kosovac

Our lives are currently being upended as a result of the Covid-19 pandemic. The bulk of public discussion is focused on 'when can we go back to normal?', 'normal' being, of course, the way we lived before the onset of lockdowns and strict distancing measures. However, many now recognize that our previous normal cannot exist again. Hartley and Kuecker (2020) find a parallel for this thinking in water management, the 'normal' being our convenient and indulgent consumptive habits as one of the 'moral hazards' of smart water management (SWM). Resource optimism – the idea that technology will always be there to save us – is a pervasive notion that negates the need for collective responsibility and behavioural change. As Hartley and Kuecker contend, it only buys us more time until we have to face reality; in effect, it is the representation of the kicking of the intergenerational can down the road.

Technocratic approaches to water, Hartley and Kuecker argue, represent much of the problem. The recent International Water Resources Association (IWRA) report on SWM by Hartley and Kuecker demonstrates the large number of projects employing SWM in practice. The report assesses 19 projects, and all are technological in nature, with a high prevalence of sensor-driven, real-time modelling and gadgetry. This approach is not surprising, considering governments are increasingly moving to 'data-driven' decision-making, without a clear understanding of the inherent biased nature of metrics. The 'objectiveness' that comes from attributing figures to potential outcomes is a flawed, albeit alluring, system. Risk matrices are classic examples of these. Despite the risk literature having moved drastically away from traditional likelihood and consequence matrices, these still pervade much of water decision-making today (Kosovac et al., 2019). It is the pure magnetism of technocracy.

An area not explored by Hartley and Kuecker (2020), but well aligned to their argument, is the feminist contribution to the debate. As the overarching technocratic narrative represents one of 'control', such as rerouting waterways and 'creating' new water for use through updated treatment technologies, one cannot ignore the undercurrents of hypermasculine identities of power pervading this field (Vera Delgado & Zwarteveen, 2017). For too long, the feminist input into water engineering has relied predominantly on increasing the representation of women in water management, with little understanding of how the underlying structure is intrinsically masculine (Zwarteveen, 2008). An example is the endearing SWM approach, which essentially acts as an extension of this existing symbolism and identity of 'hegemonic masculinity' (Connell & Messerschmidt, 2005) pervasive in water engineering. Autocratic power as a feature of such a system reveals the ongoing struggle to harness and control natural environments, with hegemonic and nostalgic notions of modernization still firmly intact. Power is the key motif in such environments. What can SWM offer feminism apart from continuing an age-old paradigm of perceived rationality through technocratic, power-hungry means? Very little, I would argue.

Instead, I ask whether there is a greater role for feminist rhetoric to play in restructuring the relationships we have with water, and also questioning the structural power that we, as water managers, try to impose on the resource. Masculinity is well acknowledged to be an obsessively technology-focused identity (Lohan & Faulkner, 2004) that disproportionately influences water management and steers it towards the power-hungry motifs that serve to reinforce hegemonic gender norms and ideals. The preoccupation with infrastructure and technological advancements goes to the heart of Beck's (1992) notion of the risk society, creating solutions for our own self-imposed risks, while in turn generating new risks through this process. This is true of projects that introduce inter-basin transfers, of desalination, and of other areas of technocratic decision-making that favour infrastructure solutions to solve the wicked problems of today, in turn creating their own social and environmental problems. Could the imposed structural masculinity in SWM be creating larger risks than those that we are attempting to mitigate?

Progress is being made in some areas. In Australia, there are attempts to incorporate other forms of knowledge and attitudes, such as that of traditional owners, to the relationship with water as a way of sustainably managing the resource into the future. New Zealand is also currently incorporating indigenous knowledge into their own disaster risk-management processes. Strides have been made in environmental law, in understanding water as a living being in the eyes of the law: essentially, that a waterway be granted legal personhood. There are many differing approaches to water management being undertaken that are not sufficiently addressed by SWM, integrated water management or the water-sensitive cities model. These new narratives are not about taking the masculine preoccupation with 'control', but rather attempting to remove the power dynamic to work together with a resource that is essential to our existence.

Feminist and post-colonial framings therefore can provide a valuable insight into critiquing current models of water governance to help us question the overreliance on technology-driven approaches to managing water, or whether there may be opportunities for social or fringe solutions to be incorporated into the water management mix. These framings buttress, rather than weaken, the arguments advanced by Hartley and Kuecker (2020) on the dangers of technocracy. The embracing of this new perspective provides the impetus needed to reconfigure our relationship to water, away from one that is paternalistic to an approach devoid of control-based obsession that works ultimately to serve the environment and, in turn, us.

As we are full swing into one of history's worst pandemics, we start to question the level of collective action that can be relied upon to reduce our future exposure to risks. More than ever, social alternative solutions, based on collective action and away from pure technology, need to be adopted, and critically so in the field of water management.

Acknowledgments

The author thanks Kris Hartley and Glen Kuecker for a thought-provoking article; and Dan Pejic and Raya Marina Stephan for reviews of the paper.

Disclosure statement

No competing interests to declare.

ORCID

Anna Kosovac http://orcid.org/0000-0003-1845-2622

References

Beck, U. (1992). *Risk society: Towards a new modernity*. Sage.
Connell, R. W., & Messerschmidt, J. W. (2005). Hegemonic masculinity. *Gender & Society*, *19*(6), 829–859. https://doi.org/10.1177/0891243205278639
Hartley, K., & Kuecker, G. (2020). The moral hazards of smart water management. *Water International*, *45*(6), 693–701. https://doi.org/10.1080/02508060.2020.1805579
Kosovac, A., Davidson, B., & Malano, H. (2019). Are we objective? A study into the effectiveness of risk measurement in the water industry. *Sustainability*, *11*(5), 1279. https://doi.org/10.3390/su11051279
Lohan, M., & Faulkner, W. (2004). Masculinities and technologies. *Men and Masculinities*, *6*(4), 319–329. https://doi.org/10.1177/1097184x03260956
Vera Delgado, J. R., & Zwarteveen, M. (2017). Queering engineers? Using history to re-think the associations between masculinity and irrigation engineering in Peru. *Engineering Studies*, *9*(2), 140–160. https://doi.org/10.1080/19378629.2017.1361427
Zwarteveen, M. (2008). Men, masculinities and water powers in irrigation. *Water Alternatives*, *1* (1), 111–130.

Deconstructing masculinity in water governance

Kris Hartley and Glen Kuecker

Introduction: illusions of progress in water management

The 21st century is a moment of reckoning for the field of public policy. Multiple convergent crises threaten ecological, economic and social stability while testing the planning and response capacities of the public sector. Legacy models of governing, including democracy, are strained by the rise of illiberal political movements built on regressive dogmas and authoritarian tactics. Geopolitical tensions simmer as global actors claim physical and ideological territory. Amidst these forces, scholarship on public policy – while lively and engaged – struggles to influence practice and is often reduced to theoretical navel-gazing. Indeed, there are many challenges to be addressed if humanity expects to survive the coming period of existential turbulence. Connecting theory and practice can help reveal new ways of understanding and approaching policy problems.

The water sector is one arena in which such turbulence is playing out. As a field historically dominated by rationalist perspectives and engineering-heavy solutions, water governance operates within shifting ecological, economic and socio-political settings that raise questions about the epistemic foundations of the field. We refer to epistemic foundations as the perspectives and methods by which practitioners and scholars select what to know about a policy problem and how to respond. We argue that these epistemic foundations, and indeed those across all domains of public policy, are the evolutionary product of decades of technocratic thinking that we label 'instrumental rationalism' (the '-ism' implies an underlying ideological agenda that is not necessarily implied by the term 'rationality'). In our 2020 commentary for *Water International* entitled 'The moral hazards of smart water management', we elaborate on this argument:

> Instrumental rationalism, the applied manifestation of technocracy partially responsible for exacerbating moral hazard, is the proverbial hammer that treats every problem like a nail; the broken [water] pipe is fixed without deeper reflection about why it was broken. Behaviours that broke the pipe fail to change because the fix is always available, invoking moral hazard by illustrating the 'safety' of maintaining existing habits and the externalization or socialization of costs associated with them. (Hartley & Kuecker, 2020, p. 695)

The idea that there exists a profound flaw in the longstanding epistemic orientation of water governance accords with critical–theoretical efforts to reform governance systems that tend towards oppression and exclusion. Relatedly, a provocative commentary by

Kosovac (2021), in response to our own piece quoted above, develops its own important ideas by suggesting the potential for feminist critiques to improve how water governance is understood and practiced. Feminist and post-colonial framings, according to Kosovac, 'provide a valuable insight into critiquing current models of water governance to help us question the overreliance on technology-driven approaches to managing water' (p. 343). Kosovac, correctly in our view, attributes the field's preoccupation with technocratic and engineering solutions to a masculinist–paternalist ambition to control the unpredictable and undisciplined – in this case, the natural environment. According to Wolfe and Brooks (2017), undergirding the desire for water managers to cling to this perspective are survivalist instincts and the seeming usefulness of ringfencing nature as separate from humanity; the authors state that 'mortality salience, arising from exposure to conscious and unconscious death reminders, helps explain why individuals and societies want to control water' (p. 2).

We enthusiastically embrace Kosovac's ideas about the urgency to rethink water management and concur that a feminist critique is needed in public policy's current age of epistemic destabilization. Developing feminist critiques of technocratic (or 'smart') approaches to water management is a crucial task for breaking epistemic path dependencies and transcending the Cartesian human–nature split. Critical appraisals of this sort have high potential to expose what Kosovac calls the 'hyper masculine identities of power' within smart water management. This potential is evident, for example, in efforts by water utilities to diversify their workforces. Perfunctory initiatives to hire more women, ostensibly in an effort to rectify the entrenched masculinization of water governance, reflect the folly of technocratic solutionism. Hiring more women may help meet a water utility's personnel quota but does little to proactively confront an underlying mindset that bears the legacy of a masculinist rationality seeking to conquer and subdue nature. Like efforts to pipe away wastewater, hiring a higher percentage of women is an immediately measurable solution about which water managers can boast in the short-run – excusing them from the more difficult qualitative work of systemic and cultural reform. These perspectives reflect what Wolfe and Brooks (2017) identify as a 'distal defense' in which powerful actors valorize 'individual and group worldviews and [pursue] culturally sanctioned hero projects to maximize individual self-esteem' (p. 5).

Positioning the feminist critique on water management

We recognize the profound challenge that Kosovac's proposal makes to the field of water management – and by extension to multiple domains of public policy. In the remainder of this article, we lend our own perspectives to Kosovac's ideas. In particular, we identify two issues deserving more attention: one is familiar to feminist theory, while the second is perhaps novel. The first pertains to which feminist critique the field of water management can most productively engage. Kosovac makes clear that the masculinization of water management is nothing that a simple hiring initiative can alone correct; we acknowledge the value of such an initiative while maintaining a realistic perspective about the difficulty of reshaping professional culture and epistemic orientations. Furthermore, efforts at gender balance in hiring largely ignore the more revolutionary call by feminist theory for fundamental reform. The uncompleted task, then, is for scholars (and practitioners, where possible) to develop lines of critical enquiry and analysis that begin with and are driven by the manifold tenets underlying feminist theory – namely those related to power inequalities in all social units, from families to organizations to societies. The mechanics of how gender inequalities are understood can be applied also to studies about inequalities across multiple types of difference including race, class and other dimensions of personal identity.

In illustrating how masculinist power dynamics replicate themselves in even subtle ways, we select our own previously quoted article as an example. We acknowledge first that we, the authors, are both male. The implication of this acknowledgement is that while we seek to engage with feminist theory in this article (as we do critical theory in our other writings) as a way to bring deeper understanding to governance problems, we do not have the direct experience with gender discrimination that would imbue our writing with personal authenticity. This is evident both in what we wrote and, more importantly, what we did not write in the aforementioned article. Our analysis did not recognize the potential contribution of feminist theory, nor did it make even a tangential reference to it. We could claim that this is a professional oversight or that our theoretical framing was ringfenced in a way that excused us from surveying the entire landscape of critically oriented perspectives. Yet, viewed more reflectively, our oversight is a type of discursive silencing that, if not fully embracing masculinist perspectives, did nothing to explicitly name and confront them. Work such as ours risks reproducing the patriarchy of water management by overlooking feminist theory – an oversight that, regardless of our intent, has a marginalizing effect. Our arguments in the current article recognize that our claims, even in critique of our own work, bear a perspective that is shaped by our experiences as men working in a profession that studies water governance. In extending our reflection on this experience, we are reminded of our own work on the processes by which the path dependency of 'received wisdom' (i.e., that shaped by hegemonic actors) leads to a 'common sense' epistemic that tolerates no dissent (Hartley & Kuecker, 2021). The self-citation we made in the previous sentence (and have elsewhere in this article) could also be seen as an example of the type of rhetorical mechanism that allows hegemonic common sense to auto-replicate and protect itself from meaningful critique. Masculinist perspectives related to the rationalizing and controlling approach of technocratic or smart water management are, we argue, a mere construct that make illegitimate claims on common sense but enjoy the privilege of that status nonetheless.

In closing this section's argument about the applicability of feminist theory to the analysis and practice of water management, we return to the claim that recognizing the need for feminist theory is one thing, but discerning which feminist theory is another. In this instance, the 'problem' of post-colonial theory arises; as Mohanty (2003) shows in *Feminism Without Borders*, there exists a need – if not an imperative – to decolonize feminist theory so that the diversity of perspectives is not overwhelmed or co-opted by the dominant discourse generated by Western feminist activists and scholars. Decolonizing knowledge, however, is a fraught undertaking that interrogates epistemic foundations while lacking a consensus methodological or theoretical approach. Even the scope of eligible targets is unclear; Tuck and Yang (2012) caution against extending 'decolonization' beyond critiques of settler colonialism's displacements of the indigenous. For example, research by Speed et al. (2006) highlights the conflictual terrain within the Zapatista revolution (Mexico), where the *normas y costumbres* of indigenous culture, supporting an autonomous political movement that asserts indigenous people's power to control their destiny, had to engage with Western feminist proposals and practices that were exogenous to their culture. As feminist theory raises the necessary questions of 'whose smart water' and 'for whose agenda', we need also to ask similar of feminist theory ('whose is privileged') – lest the promise of liberation made by post-colonial and critical theory be compromised.

Feminist perspectives to escape anachronistic public policy

The second issue emerging from Kosovac's article deserving further attention is the positioning of water management within the exigencies of the (currently ongoing) global Covid-19 pandemic and, in particular, how an acute convergence of multiple 'wicked' problems and crises presages epistemic rupture. We embrace the implication made by Kosovac and other scholars that humanity will not return to its pre-pandemic norm; following Roy's (2020) pandemic metaphor, humanity has entered a portal leading to an uncertain future. The prospect of a permanently altered condition points to the wicked predicaments of a pandemic-driven systemic collapse, which Kuecker (2020) argues has led humanity to a liminal state where the epistemic destabilization of modernity's truth claims unleashes an era of disruption and rupture. The outcome of the liminal state is impossible to anticipate and define before it unfolds because it is an emergent property freed from the structural constraints of pre-pandemic (or, more abstractly, pre-disruption) thought systems that made technocratic rationalism hegemonic.

As Hartley et al. (2019) argued with respect to policymaking theories, legacy ways of thinking become anachronistic in the face of epistemic disruption; the disruption at-hand is the prospect of systemic collapse that was already underway before the pandemic. In the liminal state, legacy knowledges maintain their credibility and remain influential even amidst the emergence of alternative perspectives (however contested). Accordingly, there is productive potential in positioning feminist theory within the concept of the liminal state, in that feminist theory represents one epistemic (among many) that passes through Roy's existential portal to constitute the thought system emerging from the ashes of 'soft collapse'.[1] At the same time, anachronistic epistemics in their privileged 'legacy' status are positioned to regain their footing and colonize, co-opt or marginalize what emerges during a period when all knowledge is questioned. From a practical perspective, this epistemic noise reflects what Wolfe and Brooks (2017) describe as a 'water soft path' (p. 9) in which past trends and ways of thinking are abandoned in favour of a 'backcasting' approach that begins with a vision of the future and derives a pathway forward therefrom.

It is prudent to recognize that feminist perspectives sit among a constellation of perspectives – both 'critical' and otherwise – that will vie for privileged stead in a liminal period. The truth claims of modernity are being challenged, but not only by critical theorists. So-called 'anti-vaxx' movements, efforts to discredit democratic electoral integrity, scepticism about climate science and challenges to the legitimacy of institutions such as legislatures and the judiciary have recently enthralled the political 'right' in many countries, including the United States and UK. At a higher conceptual level, these movements also reflect the disruptive power of contrarian ideologies across the political spectrum. In this way, challenging the epistemic foundations of policy knowledge can be seen as both a theoretical exercise and a practical political tool. Addressing knowledge contestation in reference to climate change denial, Fischer (2019) states:

> what the deniers are up to is offering a counter-narrative – based on what they see as the alternative facts of an alternative reality – that challenges this translation [of scientific knowledge to political knowledge]. One does not have to accept their arguments, rather only to recognize that this is how political knowledge is generated and that the arguments legitimately follows [*sic*] the logic of practical sociopolitical argumentation. (p. 148)

Denialist and contrarian political phenomena reflect the type of 'negotiation' (however disingenuous) of truths and narratives that characterize discourse in democratic settings. As such, challenging the epistemic foundations of policy narratives is not a fundamentally left- or right-wing exercise; it is one rhetorical mechanism or analytical perspective among many. At the same time, we maintain that reframing of knowledge can occur at a meta-level that is not beholden to the dictates of ideology but exhibits how even shared understandings about policy are based on deeply embedded assumptions (i.e., a masculinist preoccupation with control) that have long enjoyed the privilege of eluding meaningful critical scrutiny.

Conclusions: bridging theory and practice

The roots of deep systemic faults are often obscured by the banality of 'business as usual'. Do the ideas of critical theorists keep the lights on or the water running? Shaming theory for its detached abstraction and alleged irrelevance to practice is a near universal indulgence – if also trite. Nevertheless, Kosovac's effort to bridge the gap between a highly 'practical' policy domain (public utilities) and a decades-long academic discussion about power imbalances challenges scholars and practitioners to reach beyond what their education and professional socialization (*habitus*) have conditioned them to understand. The call to feminize critiques of water recognizes the role of power and struggle not only in the mechanics of public management but also, more sublimely, in the political economy of knowledge and the process by which understandings about policy problems are framed to serve elite interests. The latter suggests a need to look beyond the tropes of interdisciplinarity or applied research and anticipate the next great moment in human understanding – a new 'Enlightenment'.[2] Within the tools of technocratic and smart water management there exists no potential for Enlightenment but only a doubling-down on legacy ways of approaching policy problems. In short, modern water management is a study in how to do the wrong things, but faster and cheaper. Through the proverbial emperor's clothes of the 'smart' label, technocratic rationalism is poised to reproduce itself within the liminal state of epistemic transition – it will promise novelty while delivering only repackaged dross. Feminist theory can intervene by exposing and deposing the masculinist epistemic of command and control, and by proposing a bold vision for how transcendent understandings about public policy can emerge.

Notes

1. Hartley and Kuecker (2021) define a soft collapse as follows: 'A system's movement from ordered complexity towards disordered simplicity, during which the integrity of system complexity is not entirely lost while the system, to some extent, enters a liminal state with an uncertain outcome. During a soft collapse, failures of the hegemonic epistemic and its policy manifestations produce observable systemic destabilization and decline. Defenders of the hegemonic epistemic respond by doubling-down on legacy modes of thinking while defending their credibility against an onslaught of anomalous data and political pushback. The doubling-down results in momentary but illusory resolution of policy problems without

progress on underlying determinants. The temporarily successful state ultimately gives way to an eventual hard collapse.'

2. We acknowledge that the term 'Enlightenment' is a Western construct and that the evolution of knowledge, however defined, occurs in parallel across societies over time. The epistemic orientation that emerged from the Western Enlightenment, as understood according to our arguments in this article, gained its hegemony during an historical moment that saw a particular convergence of global power dynamics. This convergence and the global 'order' that resulted had the effect of privileging scholarly work and cultural understandings that originated in Western countries. As such, we use the term 'Enlightenment' generically in reference to a moment of epistemic reckoning that renders existing knowledge anachronistic.

Disclosure statement

No potential conflict of interest was reported by the authors.

ORCID

Kris Hartley http://orcid.org/0000-0001-5349-0427
Glen Kuecker http://orcid.org/0000-0001-5522-2821

References

Fischer, F. (2019). Knowledge politics and post-truth in climate denial: On the social construction of alternative facts. *Critical Policy Studies*, *13*(2), 133–152. https://doi.org/10.1080/19460171.2019.1602067

Hartley, K., & Kuecker, G. (2020). The moral hazards of smart water management. *Water International*, *45*(6), 693–701. https://doi.org/10.1080/02508060.2020.1805579

Hartley, K., & Kuecker, G. (2021). *Disrupted governance: Towards a new policy science*. Cambridge University Press.

Hartley, K., Kuecker, G. D., & Woo, J. J. (2019). Practicing public policy in an age of disruption. *Policy Design and Practice*, *2*(2), 163–181. https://doi.org/10.1080/25741292.2019.1622276

Kosovac, A. (2021). Masculinity and smart water management: Why we need a critical perspective. *Water International*, *46*(3), 342–344. https://doi.org/10.1080/02508060.2021.1886832

Kuecker, G. D. (2020, December). The perfect storm's pandemic driven soft collapse. *The International Journal of Environmental, Cultural, Economic and Social Sustainability*, *16*(1), 1–18 https://doi.org/10.18848/1832-2077/CGP/v16i01/1-18.

Mohanty, C. T. (2003). *Feminism without borders: Decolonizing theory, practicing solidarity*. Duke University Press.

Roy, A. (2020, April 3). The pandemic is a portal. *Financial Times*. https://www.ft.com/content/10d8f5e8-74eb-11ea-95fe-fcd274e920ca.

Speed, S., Castillo, R. A. H., & Stephen, L. (Eds.). (2006). *Dissident women: Gender and cultural politics in Chiapas*. University of Texas Press.

Tuck, E., & Yang, K. W. (2012). Decolonization is not a metaphor. *Decolonization: Indigeneity, Education & Society*, *1*(1), 1–40. https://jps.library.utoronto.ca/index.php/des/article/view/18630

Wolfe, S. E., & Brooks, D. B. (2017). Mortality awareness and water decisions: A social psychological analysis of supply-management, demand-management and soft-path paradigms. *Water International*, *42*(1), 1–17. https://doi.org/10.1080/02508060.2016.1248093

CONCLUSION

Before you go: the editors' checklist of what we now know about Smart Water Management

Stephanie Kuisma, James E. Nickum, Henning Bjornlund and Raya Marina Stephan

We have created the following checklists of key points about what we now know, as well as the hopes and fears about Smart Water Management (SWM), based on our reading of the materials in this special issue. You may have others to add. This is just the beginning of the discussion. IWRA is staying engaged on this critical issue. There will be more to come, whether in the pages of *Water International*, in the Policy Briefs series (www.iwra.org/policybriefs), or in co-sponsored activities and reports with IWRA.

SWM promises to

- Help improve access to and affordability of drinking water
- Improve the flexibility, efficiency and reliability of water systems
- Increase knowledge and decision-making opportunities through the ability to diagnose, inspect and repair problems in water infrastructure when they arise
- Improve data quality, access, consistency, transparency and frequency of observation
- Manage/reduce financial risk and/or improve the financial position of water utilities through reduced infrastructure costs, improved capacity, and reduced water loss
- Improve the quality of service provided by water utilities, including increased trust in water suppliers through transparency
- Increase health and safety through improved water, sanitation and hygiene (WASH)
- Provide the opportunity to test, develop and leapfrog over past water system challenges
- Improve water for all, including the environment, through improved water quality and quantity, reduced groundwater depletion, energy conservation, and flood and drought management
- Enable advanced solutions tailored to local conditions
- Support low-cost shared solutions through adoption of open source technology
- Bring accessibility and affordability to remote areas
- Support learning, behavioural change, community-led decision making, and training and capacity building at all scales, even the very local
- Empower and engage marginalized communities.

Those promises can be dashed by

- Technology being used to patch up a broken or outdated system rather than address fundamental problems
- SWM being adopted without imperative critical thinking to avoid major risks

- Ignoring more promising alternative narratives and options
- Locking society into falsely thinking that technology will solve all water problems
- Not addressing underlying issues, such as continuing use of potable water for non-potable uses
- Not recognizing that posited long-term benefits remain speculative
- Security and privacy concerns
- Limiting our current thinking to the technological solutions on offer
- Impacts on the labour force of water utilities
- Being locked into a technology, when technology is advancing rapidly
- Not adapting to the local context and needs
- Moving too quickly and not taking the necessary iterative approach to integration
- New data impacting water access for transboundary waters.

The policy messages: what needs to be done

- Think critically of alternative narratives: what could go wrong? – it probably will.
- Design guiding principles in consultation with all affected parties.
- Operate with a long-term vision, including support for long-term investments for adequate research, development, testing and implementation.
- Tailor solutions to different contexts.
- Reassess what our water systems should look like, instead of fixing broken systems.
- Support municipalities to design better cities with water in mind.
- Coordinate efforts with community, leaders and experts.
- Generate a stronger political commitment to fix the system with policies and governance as well as technologies.
- Complement technical SWM with innovative smart institutional management.
- Ensure that the technologies adopted are robust, easy to use, adaptive, and appropriate to each context, and can communicate with each other
- Ensure that all benefits are considered when developing SWM solutions, including social and environmental benefits, as well as revenue streams.
- Think outside the silo, including consideration of the full hydrological cycle.

Stephanie Kuisma

James E. Nickum

editor@iwra.org

Henning Bjornlund

http://orcid.org/0000-0003-3341-5635

Raya Marina Stephan

Index

Page numbers in **bold** refer to tables and those in *italic* refer to figures.

For Product Safety Concerns and Information please contact our EU
representative GPSR@taylorandfrancis.com
Taylor & Francis Verlag GmbH, Kaufingerstraße 24, 80331 München, Germany

www.ingramcontent.com/pod-product-compliance
Ingram Content Group UK Ltd.
Pitfield, Milton Keynes, MK11 3LW, UK
UKHW050054200926
475702UK00019B/2834
9781032345741